PRAISE FOR *SI*  G

"There are things in this book that cou._____ ____ ____ your head."

—Vernor Vinge, computer scientist; Hugo Award-winning author, A Fire
Upon the Deep; *essayist, "The Coming Technological Singularity"*

*"The arrow of progress may kick upwards into a booming curve or it may terminate
in an existential zero. What it will not do is carry on as before. With great insight
and forethought, Miller's* Singularity Rising *prepares us for the forking paths ahead
by teasing out the consequences of an artificial intelligence explosion and by staking
red flags on the important technological problems of the next three decades."*

—Peter Thiel, self-made technology billionaire; co-founder,
Singularity Summit

"Many books are fun and interesting, but Singularity Rising *is fun and interesting
while focusing on some of the most important pieces of humanity's most important
problem."*

—Luke Muehlhauser, executive director, Singularity Institute

*"We've waited too long for a thorough, articulate, general-audience account of mod-
ern thinking on exponentially increasing machine intelligence and its risks and
rewards for humanity. Miller provides exactly that, and I hope and expect that his
book will greatly raise the quality of debate and research in this critical area."*

—Aubrey de Grey, leading biomedical gerontologist; former AI researcher

*"How can we be intelligent about superintelligence? Its finessed agility steers its
course through the terrain of analytics and into the salty basin of awareness. It is
wise. It is a nonpartisan player. It flirts freely with friendliness. Miller understands
this, even if his approach is at times jolting.* Singularity Rising, *by default, turns
the reader to question the true value of intelligence and hopefully realize that it must
be found in the bosom of its wisdom."*

—Natasha Vita-More, chairman, Humanity+;
editor, The Transhumanist Reader

SINGULARITY RISING
SURVIVING AND THRIVING IN A SMARTER, RICHER, AND MORE DANGEROUS WORLD

James D. Miller

BenBella Books, Inc.
Dallas, TX

Copyright © 2012 by James D. Miller

BenBella Books, Inc.
10300 N. Central Expressway
Suite #400
Dallas, TX 75231
www.benbellabooks.com
Send feedback to feedback@benbellabooks.com

Printed in the United States of America
10 9 8 7 6 5 4 3 2 1

Library of Congress Cataloging-in-Publication Data is available for this title.

ISBN 978-1-936661-65-7

Editing by Erin Kelley
Copyediting by Annie Gottlieb
Proofreading by Amy Zarkos and Cape Cod Compositors, Inc.
Indexing by Clive Pyne
Cover design by Sarah Dombrowsky and Moxie Studio
Text design and composition by RepoCat LLC
Printed by Berryville Graphics

Distributed by Perseus Distribution
perseusdistribution.com

To place orders through Perseus Distribution:
Tel: 800-343-4499
Fax: 800-351-5073
E-mail: orderentry@perseusbooks.com

Significant discounts for bulk sales are available. Please contact Glenn Yeffeth at glenn@benbellabooks.com or 214-750-3628.

I dedicate this book to those working to bring about a positive
Singularity, particularly the employees and financial backers of the
Singularity Institute for Artificial Intelligence.
On your labors might rest the future of mankind.

Contents

INTRODUCTION   IX

PART 1 RISE OF THE ROBOTS
CHAPTER 1 Exponentially Improving Hardware   3
CHAPTER 2 Where Might the Software Come From?   7
CHAPTER 3 Unfriendly AI Terrifies Me   21
CHAPTER 4 A Friendly Explosion   35
CHAPTER 5 Military Death Race   47
CHAPTER 6 Businesses' AI Race   55

PART 2 WE BECOME SMARTER, EVEN WITHOUT AI
CHAPTER 7 What IQ Tells You   63
CHAPTER 8 Evolution and Past Intelligence Enhancements   75
CHAPTER 9 Increasing IQ Through Genetic Manipulation   83
CHAPTER 10 Cognitive-Enhancing Drugs   101
CHAPTER 11 Brain Training   113
CHAPTER 12 International Competition in Human-Intelligence Enhancements   119

PART 3 ECONOMIC IMPLICATIONS
CHAPTER 13 Making Us Obsolete?   131
CHAPTER 14 How Cognitive-Enhancing Drugs Might Impact the Economy   155
CHAPTER 15 Inequality Falling   165
CHAPTER 16 Preparing for the Singularity   175
CHAPTER 17 What Might Derail the Singularity?   197
CHAPTER 18 Singularity Watch   209

ACKNOWLEDGMENTS   223

NOTES   225

REFERENCES   235

INDEX   245

INTRODUCTION

We are on the edge of change comparable to the rise of human life on Earth.

—*Vernor Vinge*[1]

Economic prosperity comes from human intelligence. Consider some of the most basic human inventions—the wheel, the alphabet, the printing press—and later, more complex and advanced inventions such as indoor plumbing, automobiles, radio, television, and vaccines. All are products of the human brain. Had our species been a bit less bright, these inventions might have escaped us. Yet we can only begin to imagine the many additional wondrous technologies we might now possess had evolution made us even smarter.

In the past, human intelligence was a gift of evolution. No more. We are now using our intelligence to figure out ways of increasing our brainpower.

The rapidly falling cost of gene sequencing will soon let us unlock the genetic basis of intelligence. Combining this knowledge with already existing fertility treatments will allow parents to raise the average intelligence of their children, while merging this genetic data with future reproductive technologies might yield children smarter than have ever existed. Even if democratic countries reject these biotechnologies, the historically pro-eugenic Chinese probably won't. As I predicted in 2007,[2] China has already embarked on a program to identify some of the genes behind genius.[3]

Artificial intelligence (AI) offers another path to expanding the sum of intelligence available to mankind. Over the coming decades, scientists may take advantage of continuous exponential improvements in computing hardware either to create stand-alone, general-purpose machine intelligences or to integrate AI into our own brains.

Vast increases in biological and machine intelligences will create what's being called the *Singularity*—a threshold of time at which AIs that are at least as smart as humans, and/or augmented human intelligence, radically remake civilization.

A belief in a coming Singularity is slowly gaining traction among the technological elite. As the *New York Times* reported in 2010, "Some of Silicon Valley's smartest and wealthiest people have embraced the Singularity."[4] These early adopters include two self-made billionaires: Peter Thiel, a financial backer of the Singularity Institute for Artificial Intelligence, and Larry Page, who helped found Singularity University. Peter Thiel was one of the founders of PayPal, and after selling the site to eBay, he used some of his money to become the key early investor in Facebook. Larry Page cofounded Google. Thiel and Page obtained their riches by successfully betting on technology.

Famed physicist Stephen Hawking is so concerned about a bad Singularity-like event that he warned that computers might become so intelligent that they could "take over the world." Hawking also told the president of the United States that "unless we have a totalitarian world order, someone will design improved humans somewhere."[5]

FIVE UNDISPUTED FACTS THAT SUPPORT THE LIKELIHOOD OF THE SINGULARITY

1. **Rocks exist!**

   Strange as it seems, the existence of rocks actually provides us with evidence that it is possible to build computers powerful enough to take us to a Singularity. There are around 10 trillion trillion atoms in a 1-kilogram (2.2-pound) rock, and as inventor and leading Singularity scholar Ray Kurzweil writes:

   > Despite the apparent solidity of the object, the atoms are all in motion, sharing electrons back and forth, changing particle spins, and generating rapidly moving electromagnetic fields. All of this activity represents computation, even if not very meaningfully organized.[6]

   Although we don't yet have the technology to do this, Kurzweil says that if the particles in the rock were organized in a more "purposeful manner," it would be possible to create a computer trillions of times more computationally powerful than all the human brains on Earth combined.[7] Our eventual capacity to accomplish this is established by our second fact.

2. **Biological cells exist![8]**
The human body makes use of tiny biological machines to create and repair cells. Once mankind masters a similar kind of nanotechnology, we will be able to cheaply create powerful molecular computers. Our third fact proves that these computers could be turned into general-purpose thinking machines.

3. **Human brains exist!**
Suppose this book claimed that scientists would soon build a human teleportation device. Given that many past predictions of scientific miracles—such as cheap fusion power, flying cars, or a cure for cancer—have come up short, you would rightly be suspicious of my teleportation prediction. But my credibility would jump if I discovered a species of apes that had the inborn ability to instantly transport themselves across great distances.

In some alternate universe that had different laws of physics, it's perfectly possible that intelligent machines couldn't be created. But human brains provide absolute proof that our universe allows the construction of intelligent, self-aware machines. And, because the brain exists already, scientists can probe, dissect, scan, and interrogate it. We're even beginning to understand the brain's DNA- and protein-based "source code." Also, many of the tools used to study the brain have been getting exponentially more powerful, which explains why engineers might be within a couple of decades of building a working digital model of the brain, even though today we seem far from understanding all of the brain's operations. Would-be creators of AI are already using neuroscience research to help them create machine-learning software.[9]

Our fourth fact shows the fantastic potential of AI.

4. **John von Neumann existed!**
It's extremely unlikely that the chaotic forces of evolution just happened to stumble on the best possible recipe for intelligence when they created our brains, especially since our brains have many constraints imposed on them by biology: they must run on energy obtained from mere food, must fit in a small space, and can't use useful materials, such as metals and plastics, that engineers employ all the time.

We share about 98 percent of our genes with some primates, but that 2 percent difference was enough to produce creatures that can assemble spaceships, sequence genes, and build hydrogen bombs.[10] What happens when mankind takes its next step, and births life-forms who have a 2 percent genetic distance from us?

But even if people such as Albert Einstein and his almost-as-theoretically-brilliant contemporary John von Neumann had close to the highest possible level of intelligence allowed by the laws of physics, creating a few million people or machines possessing these men's brainpower would still change the world far more than the Industrial Revolution did. To understand why, let me tell you a bit about von Neumann.

Although a fantastic scientist, a pathbreaking economist, and one of the best mathematicians of the twentieth century, von Neumann also possessed fierce practical skills. He was, arguably, the creator of the modern digital computer.[11] The computer architecture he developed, now called "von Neumann architecture," lies at the heart of most computers.[12] Von Neumann's brains took him to the centers of corporate power, and he did high-level consulting work for many private businesses, including Standard Oil, for which he helped to extract more resources from dried-out wells.[13] Johnny (as his biographer often calls him in tribute to von Neumann's unpretentious nature) was described as having "the invaluable faculty of being able to take the most difficult problem, separate it into its components, whereupon everything looked brilliantly simple...."[14]

During World War II, von Neumann became the world's leading expert on explosives and used this knowledge to help build better conventional bombs, thwart German sea mines, and determine the optimal altitude for airborne detonations.[15] Johnny functioned as a human computer as a part of the Manhattan Project's efforts to create fission bombs.[16] Whereas atomic weapons developers today use computers to decipher the many mathematical equations that challenge their trade, the Manhattan Project's scientists had to rely on human intellect alone. Fortunately for them (although not for the Japanese), they had access to Johnny, perhaps the best person on Earth at doing mathematical operations quickly.[17]

Unlike many scientists, Johnny had tremendous people skills, and he put them to use after World War II, when he coordinated American defense policy among nuclear weapons scientists and the military. Johnny became an especially important advisor to President Eisenhower, and for a while he was "clearly the dominant advisory figure in nuclear missilery."[18]

Johnny developed a reputation as an advocate of "first strike" attack and preemptive war because he advocated that the United States should try to stop the Soviet Union from occupying Eastern Europe. When critics pointed out that such resistance might cause a war, Johnny said, "If we are going to have to risk war, it will be better to risk it while we have the A-bomb and they don't."[19]

After Stalin acquired atomic weapons, Johnny helped put deadly deterrents in place to prevent Stalin from wanting to start another war. By the dawn of the atomic age, Stalin had demonstrated through his purges and terror campaigns that he placed little value on the lives of ordinary Russians. Von Neumann made Stalin unwilling to risk war because von Neumann shaped U.S. weapons policy—in part by pushing the United States to develop hydrogen bombs—to let Stalin know that the only human life Stalin actually valued would almost certainly perish in World War III.[20]

Johnny helped develop a superweapon, played a key role in integrating it into his nation's military, advocated that it be used, and then made sure that his nation's enemies knew that in a nuclear war they would be personally struck by this superweapon. John von Neumann could himself reasonably be considered the most powerful weapon ever to rest on American soil.

Now consider the strategic implications if the Chinese high-tech sector and military acquired a million computers with the brilliance of John von Neumann, or if, through genetic manipulation, they produced a few thousand von Neumann-ish minds every year. Contemplate the magnitude of the resources the US military would pour into artificial intelligence if it thought that a multitude of digital or biological von Neumanns would someday power the Chinese economy and military. The economic and martial advantages of having a von Neumann–or–above-level intellect are so enormous that if it proves practical to mass-produce them, they will be mass-produced. A biographer of John von Neumann wrote, "The cheapest way to make the world richer would be to get lots of his like."[21] A world with a million Johnnies, cooperating and competing with each other, has a reasonable chance of giving us something spectacular, beyond what even science fiction authors can imagine—at least if mankind survives the experience. Von Neumann's existence highlights the tremendous variance in human intelligence, and so illuminates the minimum potential gains of simply raising a new generation's intelligence to the maximum of what our species' current phenotype can sustain.

John von Neumann and a few other Hungarian scientists who immigrated to the United States were jokingly called "Martians" because of their strange accents and seemingly superhuman intelligence.[22] If von Neumann really did have an extraterrestrial parent, whose genes arose, say, out of an advanced eugenics program that Earth couldn't hope to replicate for a million years, then I wouldn't infer from his existence that we could get many of him. But since von Neumann was (almost certainly) human, we have a good chance of making a lot more Johnnies.

One Possible Path to the Singularity: Lots of von Neumann–Level Minds

Photo Credit[23]

Before he died in 1957, von Neumann foresaw the possibility of a Singularity. Mathematician Stanislaw Ulam wrote, in reference to a conversation that he had with von Neumann:[24]

> One conversation centered on the ever accelerating progress of technology and changes in the mode of human life, which gives the appearance of approaching some essential singularity in the history of the race beyond which human affairs, as we know them, could not continue.

Von Neumann was not a modest man; he knew that he could accomplish great things, especially compared to the average mortal. I bet that when he contemplated the future destiny of mankind, von Neumann tried to think through what would happen if machines even smarter than he started shaping our species' affairs—which leads us to our fifth fact in support of a Singularity.

5. **If we were smarter, we would be smarter!**
   Becoming smarter enhances our ability to do everything, including our ability to figure out ways of becoming even smarter—because our intelligence is a reflective superpower, able to turn on itself to decipher its own workings. Consider, for example, a college student taking a focus-improving drug such as Adderall, Ritalin, or modafinil, to help her learn genetics. After graduation, this student might get a job researching the genetic basis of human intelligence, and her work might assist pharmaceutical companies in making better cognitive-enhancing drugs that will help future students acquire an even deeper understanding of genetics. Smarter scientists could invent ways of making even smarter scientists who could in turn make even smarter scientists, *ad infinitum*. Now throw the power of machine intelligence into this positive feedback loop, and we could end up at technological heights beyond our imagination.

FURTHER IMPLICATIONS OF A SINGULARITY

From the time of Alexander the Great up to that of George Washington, the lot of the average person didn't much change because there was little economic growth. On average, a man lived no better than his great-grandfathers did. But shortly after Washington's death, an industrial revolution swept England that married science to business. The Industrial Revolution was the most important turning point in history since the invention of agriculture because it created sustained economic growth arising from innovation—the creation of new and improved goods and services. Innovation, and therefore economic growth, comes from human brains.

Think of our economy as a car. Before the Industrial Revolution, the car was as likely to move backward as forward. The Industrial Revolution gave us an engine powered by human brains. Technologies that increase human intelligence could supercharge this engine. Artificial intelligence that is beyond the ordinary human level could move our economy out of its human-brain-powered car into an AI-propelled rocket. An ultra-intelligent AI might even be able to push our economy through a wormhole to God knows where.

Let me tell you a story, soon perhaps to become true-to-life, which should put beyond all doubt the importance of the Singularity.

Imagine it's the year 2029, and Intel has just made the most significant technological breakthrough in human history. The corporate tech giant has developed an AI that does independent scientific research. Over the past month, the program has written an article on computer design that describes how to marginally improve computer performance.

What, you might ask, is so spectacular about this program? It isn't the superiority of the article it produced because a human scientist would have taken only a month to do work of equivalent quality. The program, therefore, is merely as good as one of the many scientists Intel employs. Yet because the program succeeded in independently accomplishing the work of a single scientist, the program's designers believe that within a couple of decades, technological progress will allow the AI program to function a million times as fast as it does today!

Intel scientists have such tremendous hope for their program because of Moore's Law. Moore's Law, a pillar of this book, was formulated by Intel cofounder Gordon Moore and has an excellent track record of predicting increases in computer power and performance. Moore's Law implies that the quantity of computing power you could buy for a given amount of money would double about every year. Repeated doubling makes things very big, very fast. Twenty doublings yields about a millionfold increase.

Let's imagine that Intel's AI program runs on a $1 million computer. Because of Moore's Law, in twenty years a $1 million computer would run the program a million times faster. This program, remember, currently does the work of one scientist. So if the program is running on a computer one million times faster, it will accomplish the work of a million human scientists. Take a moment to think about that.

And twenty years into that future, other businesses would eagerly use Intel's program. A pharmaceutical company, for example, might buy a thousand copies of the program to replace a thousand researchers and make as much progress in one year as a thousand scientists would in a million years—and this doesn't even include the enhancements the AIs would garner from improved software.

If Intel really does create this human-level AI program in 2029, then humans may well achieve immortality by 2049. Because of this, I sometimes end my economics classes at Smith College by saying what I will soon be explaining to you: that if civilization doesn't collapse, you all have a decent chance of not dying.

Intel's breakthrough, unfortunately, wouldn't necessarily go so well for mankind, because to do stuff, you need stuff.

Regardless of your intelligence, to achieve anything you must use resources, and the more you want to do, the more resources you need. AI technologies would, at first, increase the resources available to humans, and the AIs themselves would gain resources from trading with us. But a sufficiently smart AI could accomplish anything present-day people can do and at much lower cost. If the ultra-AIs are friendly, or if we upgrade ourselves to merge with them, then these machine intelligences will probably bring us utopia. If, however, the ultra-AIs view mankind the way most people view apes—with neither love nor malice but rather with indifference—then they will take our resources for their own projects, leaving us for dead. Our fate may be determined by how well we manage to infuse friendliness into our AIs.

WHY READ A SINGULARITY BOOK BY AN ECONOMIST?

I hope that I have convinced you by this point that learning about intelligence enhancement is well worth your time. But why should you read this particular book, given that its author is an economist and not a scientist or an engineer? One reason is that I will use economic analysis to predict how probable changes in technology will affect society. For example, the theories of nineteenth-century economists David Ricardo and Thomas Malthus provide insights into whether robots might take all of our jobs (Ricardo) and why the creation of easy-to-copy emulations of human brains might throw mankind back into a horrible pre–Industrial Revolution trap (Malthus). Economics also sheds light on many less-significant economic effects of an advanced AI, such as the labor-market consequences if sexbots cause many men to forgo competing for flesh-and-blood women.

Furthermore, the economics of financial markets shows how stock prices will change along our road to the Singularity. The economics of game theory elucidates how conflict will affect the Singularity-inducing choices that militaries will make. The economic construction called the Prisoners' Dilemma establishes that rational, non-evil people might find it in their self-interest to risk propelling humanity into a Singularity even if they know that it has a high chance of annihilating mankind. Robin Hanson, one of the most influential Singularity thinkers, is an economist. Science shows us the possibilities we could achieve; economic forces determine the possibilities we do achieve.

But despite my understanding of economics, I admit that sometimes I get confused when thinking about a Singularity civilization. In the most important essay ever written on the Singularity, Vernor Vinge, a science fiction writer and former computer scientist, explained that accelerating technology was making it much harder for him to write science fiction because, as we progress toward the Singularity, "even the most radical [ideas] will quickly become commonplace."[25] Vinge has said that just as our models of physics break down when they try to comprehend what goes on at the Singularity of a black hole (its supposed point of infinite density), so might our models fail to predict what happens in an exponentially smarter world.[26] Vinge told me that in the absence of some large-scale catastrophe, he would be surprised if there isn't a Singularity by 2030.[27]

A father can much better predict how his six-year-old will behave in kindergarten than the child could predict what the father will do at work. If the future will be shaped by the actions of people much smarter than us, then to predict it, we must know how people considerably brighter than us will behave. While this is challenging, economics might make it possible. A high proportion of economic theory is based on the assumption that individuals are rational. If decision makers in the future, human or otherwise, are smarter and more rational than we are, then economic theory might actually better describe their behavior than it does our own.

Former US Secretary of Defense Donald Rumsfeld famously said:

> There are known knowns; there are things we know we know. We also know there are known unknowns; that is to say we know there are some things we do not know. But there are also unknown unknowns—the ones we don't know we don't know.[28]

The Singularity will undoubtedly deliver many unknown unknowns. But economics still has value in estimating how unknown unknowns will affect society. Much of economic behavior in capitalist countries is based on an expectation that property rights will be valuable and respected in the future. If you're saving for a retirement that you don't expect will start for at least another thirty years, you must believe that in thirty years money will still (probably) have value, your investments will not have been appropriated, and our solar system will still be inhabitable. But these three conditions hold simultaneously only under extremely special circumstances and for only a minuscule percentage of all the possible arrangements of society and configurations of molecules in our solar system. The greater the number of unknown unknowns you expect to occur, the less you should save for retirement and the fewer investments you should make—which shows that expectations of certain types of Singularity will damage the economy.

We don't know how mankind will increase its available intelligence, but as I will show, there are so many paths to doing so, and there are such incredible economic and military benefits of intelligence enhancement, that we will almost certainly create a smarter world. This book will serve as your guide to that world. We will discuss the economic forces that will drive intelligence enhancements and consider how intelligence enhancements will impact economic forces. Along the way, you will pick up some helpful advice on how to live in the age of the coming Singularity.

This book has one recommendation that, if you follow it, could radically improve your life. It's a concrete, actionable recommendation, not something like "Seek harmony through becoming one with Creation." But the recommendation is so shocking, so seemingly absurd, that if I tell you now without giving you sufficient background, you might stop reading.

PART 1
RISE OF THE ROBOTS

Exponential growth is deceptive. It starts out almost imperceptibly and then explodes with unexpected fury—unexpected, that is, if one does not take care to follow its trajectory.

—*Ray Kurzweil*[29]

CHAPTER 1
EXPONENTIALLY IMPROVING HARDWARE

If, as the title of the book by Ray Kurzweil proclaims, *The Singularity Is Near,* then why doesn't it appear so? An ancient story about the Hindu God Krishna partially illuminates the answer:[30]

> Krishna disguised himself as a mortal and challenged a king to a game of chess. The king agreed, and allowed Krishna to name the prize he would receive if he won the game. Krishna chose a seemingly modest reward requesting that upon victory the chess board be filled with rice so that the first square of the board had one grain, the second two, the third four . . . each square containing twice as much rice as the one before it. After losing the game the king initially thought that he could easily pay Krishna the promised prize, but then it became apparent to the king that he couldn't possibly satisfy his debt.

A chess board has 64 squares. Filling up the first 5 is easy, requiring only $1 + 2 + 4 + 8 + 16 = 31$ grains of rice, not enough for a small snack. Every twenty doublings, however, means about a millionfold increase in the number of grains. By the 60th square one would need a million times a million times a million grains of rice. As I said in the introduction, repeated doubling makes things very big, very fast. The yearly doubling power of Moore's Law might bring us to Singularity.

In 1992, world chess champion Gary Kasparov, as Kurzweil tells us, "scorned the pathetic state of computer chess."[31] Yet in 1997, IBM's supercomputer Deep Blue defeated Kasparov. Did his 1997 defeat prove that Kasparov was mistaken five years earlier in thinking so little of computer chess? No. Because of exponential improvements in computing speed, it's entirely possible for a narrow computer intelligence to go from a skill level well below that of the best human to having superhuman abilities in its domain.

For many, computers as smart as people seem tremendously far off, but this might be because people are not thinking exponentially. If I told you that computers had to be a million times faster before we could create a human-level AI, then your intuitive reaction might be that it won't happen for centuries. However, if Moore's Law continues to hold, then in twenty years you will be able to buy one million times as much computing power per dollar as you can today.[32] But will Moore's Law persist? Yes, but not in its current form.

Today increases in computing power come mainly from reducing the size of computers' transistors. The smaller the transistors, the closer together you can pack them, and the shorter the distance electrons have to travel.[33] Unfortunately, quantum effects limit how small transistors can get, and problems with heat dissipation limit how closely engineers can bunch them together. Improvements in computing power under the current paradigm will probably end before we could build a single computer processor fast enough to run a von Neumann-level mind.

In the 1950s, engineers increased the power of their computers primarily by shrinking their machines' vacuum tubes.[34] Around 1960, further shrinkage was no longer feasible. Fortunately, as Kurzweil writes, transistors, already being used in radios, were ready to replace vacuum tubes in computers and continue Moore's Law.

As Kurzweil documents, the exponential improvements in computer processing power have been going on for at least a century.[35] Whenever engineers run out of ways to continue these giant gains using existing technologies, another means always arises. Kurzweil believes that three-dimensional molecular computing based on nanotubes offers a likely path to continue this trend.[36] He explains that nanotubes were first synthesized in 1991 and are "made up of a hexagonal network of carbon atoms that have been rolled up to make a seamless cylinder."[37] Furthermore, nanotubes "are very small: single-wall nanotubes are only one nanometer in diameter, so they can achieve high densities."[38] Carbon nanotubes could easily supply the hardware requirements of an AI Singularity, since one cubic inch of them would, if fully developed, be "up to one hundred million times more powerful than the human brain."[39]

To prove practical, nanoelectronics would have to be self-assembling; otherwise, hardware manufacturers would need to "painstakingly" assemble each of the trillions of circuits.[40] Fortunately for the Singularity, scientists have already made some progress in working with self-assembling circuits.[41]

Other technologies could maintain Moore's Law after it becomes impractical to further shrink transistors: building three-dimensional circuits using otherwise conventional computer chip–making techniques; computing with DNA, light, or the spin of electrons; or, most tantalizingly, by constructing quantum computers.[42] Quantum computing, which takes advantage of the strange properties of the quantum world, might never work out. If it did, however, then for some types of calculation a relatively small quantum computer would vastly outperform "any conceivable nonquantum computer."[43] Quantum computers make the proverbial "lemonade out of lemons" by taking advantage of the quantum forces that will eventually halt the shrinkage of transistors.

Kurzweil predicts that by around 2020 it will cost approximately $1,000 to buy sufficient hardware to "emulate human-brain functionality."[44] By 2030 it would take about a thousand human brains to equal the power of a single $1,000 computer.[45] Even after 2030 there will be plenty of room for computers to keep expanding their lead over natural-born human brains.[46]

We have strong economic reasons for thinking that Moore's Law will continue. As electrical and computer engineering professor and Intel consultant Eric Polizzi told me, since Intel is the company most responsible for maintaining it, and Intel would be financially decimated if the price-performance ratio of its computer chips didn't continue to improve at a rapid rate, the company has a strong incentive to maintain Moore's Law.[47] Many companies produce computer chips, and Intel is only able to keep its high profit margin in the face of this competition because it keeps producing higher-quality (if more expensive) chips than almost all of its competitors. If innovation at Intel faltered, however, then the company's products would soon be no better than its rivals', and Intel would end up losing much of its market value as it became forced to compete purely on price.

The many benefits of faster computing also give Intel an incentive to innovate. Computing performance is a bottleneck for many software products, such as virtual reality, massive data analysis, and artificial intelligence. Knowing this, Intel and others realize the profits they will be able to gain by producing more-powerful computing hardware, so I'm confident that we'll continue to see exponential improvement in the performance of computers—at least until physics finally puts a stop to it.

When my son was three, my wife and I kept telling him how quickly he was growing. He agreed with our assessment and concluded that he would soon be so tall that he wouldn't fit in our car, or even our house. And had he continued to grow at the same rate, he indeed would have become too large for our house. Of course, while he soon outgrew his car seat, we never had to buy a new house to accommodate his height because (like all children's) his rate of growth eventually slowed.

Biological limits to human growth prevent people from becoming so big they cannot fit in a typical house. Similarly, the laws of physics limit how fast a computer can process information, and even Kurzweil thinks that these limits will eventually stop the exponential growth in computing power. But an MIT engineering professor estimated that, given the known laws of physics, the ultimate laptop computer could perform 10^{51} operations per second.[48] This is almost certainly more than a trillion trillion times faster than the human brain operates.[49]

Even if computers never get any faster than they are today, it should still become practical to create a von Neumann–level AI so long as computers keep getting cheaper. Rather than running an AI program on just one computer, we could run it on hundreds or even millions of them using what's known as massive parallel processing. It's much harder to program a large number of computers working in parallel than it is to program one computer that goes very fast, so AI software would be easier to write if you had one expensive but fast computer than if you had millions of cheap but relatively slow ones. However, even though it would be difficult, it must still be possible to run a von Neumann–level mind on a massive parallel processing system, because that's what the original von Neumann had. Human intelligence arises in large part from the interaction of the billions of computational units (neurons) in our brains.

Singularity-enabling computers would require a lot of memory as well as sufficient processing speed. Fortunately for those hoping for a speedy Singularity, computing memory (both random access memory and hard drive capacity) is also undergoing exponential improvements.[50]

Human intelligence is a product of information processing in a brain made of meat....
Information processing on a faster, more durable, and more flexible platform like silicon should
be able to surpass the abilities of an intelligence running on meat if we can figure out which
information processing algorithms are required for intelligence....
                                                                    —*Luke Muehlhauser*[51]

Let an ultraintelligent machine be defined as a machine that can far surpass all the intellectual
activities of any man...Since the design of machines is one of these intellectual activities, an
ultraintelligent machine could design even better machines; there would then unquestionably
be an "intelligence explosion,"....
                                                                    —*Irving John Good*[52]

CHAPTER 2
WHERE MIGHT THE SOFTWARE COME FROM?

Computing hardware, no matter how powerful, will never suffice to bring us to a Singularity. Creating AIs smart enough to revolutionize civilizations will also require the right software. Let's consider four possibilities for how this software might arise.

1. **Kurzweilian Merger**

   Ray Kurzweil believes that our brains will provide the starter software for a Singularity. Kurzweil predicts that the Singularity will come through a merger of man and machines in which we will gradually transfer our "intelligence, personality, and skills to the non-biological portion of our intelligence."[53] Although we might never reach it, economic, medical, and military incentives push us toward a "Kurzweilian merger."

   Humans enhance themselves with tools, be it with sticks that leverage our strength, glasses that focus our vision, or helmets that protect our skulls.[54] With cochlear implants for the deaf and brain stimulation implants for Parkinson's patients, we have started directly connecting tools to our brains. Kurzweil's merger prediction comes from his thinking that the steady exponential improvements in information technology and computing hardware will lead mankind to put more and more machines of increasing power into ourselves.

What Kurzweil has going for him is a most remarkable coincidence: The process of making computers faster also makes them better suited to being placed in our bodies. The room-size ENIAC computer of the 1940s that helped John von Neumann and other weapons scientists build the hydrogen bomb would have made a poor brain implant not only because of its size and weight but also because it lacked sufficient speed and memory—its computational power was minuscule compared to that of even a single human brain. A necessary step to improve computers, as it turned out, was to make them smaller; eventually they will be so tiny that billions could fit in our bodies.

The nanotechnology that will probably allow engineers to continue to improve computing hardware would also be the key enabler of a Kurzweilian merger: that is what we would use to construct and regulate the nonbiological part of ourselves. Kurzweil expects us to develop powerful nanotechnology by around 2025.[55]

Having many tiny but powerful machines in our bodies would bring numerous benefits, including keeping us young and healthy; giving us virtual reality from within our own nervous systems; allowing us to remember anything we have ever seen or experienced; letting us transfer skills and knowledge into our brains (you'd be only a quick download away from mastering Mandarin or becoming a skilled skier); boosting our capacity for abstract thought (I'll be better at math than John von Neumann ever was!); and enabling us to directly control our emotions and levels of happiness. The military advantages of having nanotech in soldiers' bodies will likely spur the US Department of Defense (DOD) to pour billions into the industry. The DOD is already researching ways of connecting brains and machines to help troops navigate battlefields.[56]

The benefits of being stronger, smarter, and healthier are so enormous that if it's possible to use machines to improve oneself, some people will. Thanks to Moore's Law, these advantages would steadily grow larger over time.

This is your brain on Moore's Law: every year, you replace your old implants with new ones that are twice as fast. The nonbiological part of your intelligence grows exponentially, and you become very smart, very fast. And if you work for a biotech company, you'll use your extra intelligence to come up with even better ways of merging mind and machine, further pushing us to Singularity.

Kurzweil forecasts that by the 2030s, most of our intelligence will be nonbiological. He sets the date for the Singularity at 2045 because he estimates that "the nonbiological intelligence created in that year will be one billion times more powerful than all human intelligence today."[57]

The exponential improvements in computing power lend considerable robustness to Kurzweil's merger prediction.[58] Imagine, for example, that Kurzweil underestimates by a factor of a thousand how much computing power will be needed to take us to a Singularity. Because of Moore's Law yearly doubling of the amount of computing power you can buy per dollar, such a thousandfold error would cause Kurzweil's predictions to be off by only a decade.

Kurzweil believes that mankind will end up using a significant percentage of the resources of the solar system, and eventually of the galaxy, to increase its own intelligence. After we have mastered nanotechnology, it will always be possible to increase our intelligence by using additional matter. Since at least some people (myself included) would desire to become as smart as the laws of physics allow, mankind would build as many computers as possible and would consequently convert much of the matter we encounter into computing hardware. Kurzweil foresees mankind colonizing the universe at almost the maximum speed allowed by the laws of physics.[59]

I'm uncertain whether bioengineers will ever be able to figure out how to make extremely smart people by integrating computers into our brains. But the possibility that this could happen is a path to the Singularity that mankind has a reasonable chance of following.

2. Whole Brain Emulation

An argument against using the brain as the basis for AI is that our brains are so complex it might take centuries for us to understand them well enough to merge them with machines. But even if we don't completely understand how the brain works, we still might be able to create machine emulations of it. Brain emulation would essentially be an "upload" of a human brain into a computer. Assuming sufficiently high fidelity in both simulation and brain scanning, the emulation would think just as the original, biological human did. For any given input, the silicon brain and the biological brain would produce the same output. In any given situation, the emulation would think, feel, and act just as the original person would have in the same circumstances.

Think of the human brain as a black box that receives information from the body's sensory organs. The brain takes these inputs, processes them, and then outputs commands to various parts of the body. If you can get the same outputs from the same information inputs, then—the argument goes—the emulation will be just as real and just as alive as the original brain. Economist and former AI researcher Robin Hanson, who has written extensively on emulations, told me that, conditional on civilization not collapsing, he's 80 percent confident that emulations will be developed sometime between 25 and 150 years from now.[60] Hanson identifies the three key technologies needed to simulate the brain:

BRAIN SCANNING—In order to upload the brain, you would have to scan it at a sufficiently high level of detail. One way to do this would be to thinly slice the brain, scan each of the slices, and then use a computer program to assemble the two-dimensional images of each slice into a three-dimensional model of the brain. Scientists have already fully scanned the brain of a mouse, but probably not at a high enough resolution to create a simulation.[61] Obviously, this slicing method couldn't be used on a living person. Eventually, non-destructive scanning technology might become powerful enough to allow for uploading that would leave the original brain intact. Kurzweil has shown that the resolution of noninvasive brain scanning has been rapidly increasing.[62] Eventually, Kurzweil predicts, we will have billions of nanobots in our brains recording real-time data on how our brains work.[63]

A key question for brain scanning is how much image detail is needed to convert the original brain scans into an emulation. I find it extremely unlikely that creating an emulation would require scanning at the atomic level because the atoms in our brains keep moving around and getting replaced, and yet the person we are today seems to be the same person we were a year ago.

UNDERSTANDING BRAIN CELLS—Our brains contain many different types of cells, and to successfully emulate a brain, we would need to know how all of them function. As with the whole brain, you can think of each individual cell as a black box that receives inputs and produces outputs. Creating emulations would probably require us to understand only which inputs produce which outputs for each type of cell, rather than having to discover their complete biochemistry.[64] Hanson claims that figuring out the input/output relationship for any one type of brain cell isn't all that difficult, but there are many types, and not that much effort is currently being expended on figuring out how each type works. In 2012, after my interview with Hanson, Microsoft co-founder Paul Allen committed to giving $500 million to the Allen Institute for Brain Science.[65] Part of the Institute's objective is to study how the brain stores, encodes, and processes information.

FAST COMPUTERS—We would need highly powerful computers to turn brain scans into a three-dimensional model of the brain and to run emulations quickly enough to think at least as fast as a person. If the exponential doubling of Moore's Law continues, my guess is that we will almost certainly have fast-enough computers by 2050.

The ease of copying emulations from computer to computer would make creating them stupendously profitable. The first people to be emulated would almost certainly be highly productive individuals who could earn tremendous salaries and could function well as emulations.[66] True, an emulation might not be able to use a human-like body—but many people, such as authors, computer programmers, and stock analysts, don't really need bodies to do their jobs. Once you have a single emulation, the cost of making each additional emulation would be equal to the cost of buying another computer that could run the emulation. Moore's Law would cause this cost to fall exponentially.

It's counterintuitive to think that we could emulate the human brain on a computer without understanding everything about how brains work. But even without knowing much about how a piece of software operates, it's certainly possible to take software that's running on one computer, make an exact copy of it, transport the copy to another computer, and then run the software on the other computer. This is easily accomplished if the two computers have the same operating system; and, through porting, you can sometimes transfer software between different types of hardware.[67]

I spent too much of my youth playing Atari video games. Although some of my old games are probably still in my parents' basement, hardware incompatibilities stop me from running any of them on modern computers. Pretend, however, that it became very important for me to figure out how to run these games on the relatively fast computer I'm using to write this book.

One approach I could take would be to study the code of each of the Atari games and translate it into a language that my current computer can understand. But a more economical method would be to create an emulator on my current machine that makes my computer look to the Atari software like an Atari console. If, say, the software code X32 caused a blue dot to appear on the Atari, while the code for the same action is Y78 on my current computer, then the emulator would translate X32 to Y78 so that the original Atari commands would work on my Intel machine. Doing this wouldn't require me to understand why a game had blue dots. Once my computer had the emulator, it could run any Atari game without my having to understand how the software worked. An emulator for the human brain, similarly, could allow the uploading of a brain by someone ignorant of most of the brain's biochemistry.

The success of whole brain emulations would, in large part, come down to how well our brains can handle small changes because the emulations would never be perfect. Human brains, however, are extremely robust to environmental stress. You could hit someone, infect his brain with parasites, raise or lower the temperature of his environment, and feed him lots of strange information, and he'd probably still be able to think pretty much the way he did before.[68] Evolution designed our brains to not go crazy when they encounter an unfamiliar environment. Consequently, there's an excellent chance that the software "essences" of our brains are robust enough that they could survive being ported to a machine. Of course, porting might introduce alterations that evolution never had a chance to protect us against, so the changes might make our brains nonfunctional. But whole brain emulation is still a path to the Singularity that could work, even if a Kurzweilian merger proves beyond the capacity of bioengineers.

If we had whole brain emulations, Moore's Law would eventually give us some kind of Singularity. Imagine we just simulated the brain of John von Neumann. If the (software adjusted) speed of computers doubled every year, then in twenty years we could run this software on computers that were a million times faster and in forty years on computers that were a trillion times faster. The innovations that a trillion John von Neumanns could discover in one year would change the world beyond our current ability to imagine.

3. **Clues from the Brain**
    Even if we never figure out how to emulate our brains or merge them with machines, clues to how our brains work could help scientists figure out how to create other kinds of human-level artificial intelligence. More than fifty thousand neuroscientists are currently studying the brain, and the tools they use are getting exponentially more powerful.[69] Shane Legg, a machine intelligence theorist and programmer, told me he spends much of his time studying neuroscience to give him insights into how to create AI.[70] He explained that the brain has been found to use learning algorithms that were, coincidentally, being used in some limited-scope AI programs. Shane told me that very few neuroscientists are interested in AI. But he believes that if machine intelligence became more respected as an academic field, many neuroscientists might switch to AI, meaning there could be a tipping point after which AI research among brain scientists explodes.

4. **Intelligence Explosion**
    The most exciting and terrifying concept in this book is the *intelligence explosion*, because through it, a single AI might go from being human-level intelligent to being something closer to a god, what I call an *ultra-AI*, in a period of weeks, days, or even hours.[71]

If an intelligence explosion occurred, it would happen through recursive self-improvement: as the AI became smarter, it would figure out more ways to make itself smarter yet. Quickly, the positive feedback loop would make the AI as intelligent as the laws of physics allow.

Here's a simple example of how it could work: somehow, a human-level AI is created that wishes to improve its cognitive powers. The AI examines its own source code, looking for ways to augment itself. After finding a potential improvement, the original AI makes a copy of itself with the enhancement added. The original AI then tests the new AI to see if the new AI is smarter than the original one. If the new AI is indeed better, it replaces the current AI. This new AI then proceeds in turn to enhance its code. Since the new AI is smarter than the old one, it finds improvements that the old AI couldn't. This process continues to repeat itself.

The goal of many members of the Singularity movement, particularly the Singularity Institute for Artificial Intelligence, is to create a "seed AI" designed to undergo an intelligence explosion and quickly grow to become a friendly ultra-AI.

An AI would have a decent chance of exploding its intelligence if it mastered some form of cheap, all-purpose nanotechnology. An AI with a command of nanotech would be able to quickly turn raw materials into computing hardware, for example by using the carbon in the atmosphere to make diamondoid microprocessors. A human-level AI wielding nanotech could exponentially grow by creating self-replicating factories. For example, the first factory could spend an hour making two other factories and one computer. Each new factory would then itself make two additional factories and another computer. This process could quickly turn much of the earth into computational hardware.

If the first human-level AI had minimal hardware requirements and could be run off the processing power of a smartphone, then little time might elapse between the first and the billionth copy of this AI. As the number of its copies increased, the collective intelligence of the AIs would become better at figuring out how to acquire additional hardware through purchase or theft.

Since computing hardware is improving at an exponential rate, the longer it takes for someone to develop AI software, the more likely it will be that an intelligence explosion will occur. Pretend, for example, that Google just wrote the code for a human-level AI, but this code must be run on a billion-dollar computer. If the AI lacked nanotech, it couldn't possibly explode its intelligence by running itself on millions of computers. In contrast, if Google doesn't invent the AI software for thirty years, at which time computers will be a billion times cheaper than they are today, then an intelligence explosion could easily occur very soon after the code is first run. An intelligence explosion might also quickly follow the creation of human-level AI if small changes could translate into huge differences in intelligence.

In most ways, computers today are not as smart as chimps. Let's imagine that AIs get steadily smarter and within two decades are brighter than chimps. How much longer will it take them to become vastly smarter than any human who has ever existed? If the following diagram is right, we would likely have a lot of warning time between when AIs reach chimp level and when they become ultra-intelligent.

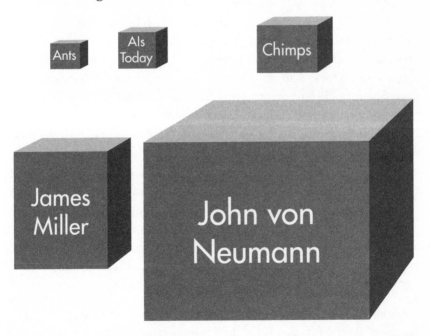

But this first diagram might overstate differences in intelligence. The basic structure of my brain is pretty close to that of chimps and shockingly similar to John von Neumann's. Perhaps, under some grand theory of intelligence, once you've reached chimp level it doesn't take all that many more tweaks to go well past John von Neumann. If this next diagram has the relative sizes of intelligences right, then it might take very little time for an AI to achieve superintelligence once it becomes as smart as a chimp.

As I'll discuss in Chapter 7, there are a huge number of genes, each of which slightly contributes to a person's intelligence. Chances are no human has come anywhere close to having all of those genes. Perhaps, after someone creates an AI based on the human brain, that AI would determine how to emulate the human brain that would arise from any given DNA. This AI could then create an emulation of a human who possessed all of the genes that enhance intelligence and none of those that retard it.

Evolution's inability to coordinate mutations provides another reason to hope—or fear—that an intelligence explosion will happen shortly after the first human-level AI is created. To grasp this, imagine that the efficiency of the human brain would be greatly increased if 100 specific mutations occurred. If it were true that the more of these mutations you had, the smarter and fitter you would be, then evolution would have an excellent chance of eventually giving our species the entire set of mutations. But if you needed to have *all* of the mutations to receive any benefit, and if having only some of them reduced the number of offspring you would have, then it is extraordinarily unlikely that the blind forces of evolution would ever produce a human possessing the full 100 mutations. In contrast, an AI might easily be able to emulate an intelligence that did have all of the mutations.

Because an intelligent AI designer of AIs could use materials, techniques, energy sources, and physical structures that evolution can't, this designer can search for brain architectures in places evolution never looked.[72] It's as if evolution gets to mine for some precious metal on Earth, whereas an intelligent AI designer can look for the metal on Earth or anywhere else in the galaxy.

Reasons to Believe in the Likely Development of AI through Either a Kurzweil Merger, Clues from the Human Brain, Brain Emulation, or Creating an AI from Scratch:

- The human brain exists, showing that it's possible to create intelligent, flexible entities that can study their own intelligence.
- The human brain runs slowly compared to what we think should be possible. As one AI theorist wrote, "Physically speaking, it ought to be

possible to run a brain at a million times the speed without shrinking it, cooling it, or invoking reversible computing or quantum computing."[73]

- The human brain is being studied by tens of thousands of neuroscientists, many of whom will undoubtedly (and at least for now, mostly unintentionally) provide clues to how to build AIs.

- Human-level or better AIs would confer significant benefits on a nation's military.

- Long-term and well-studied trends support the probable continuation of Moore's Law.

- Future technologies, such as carbon nanotubes, will likely make it possible for companies like Intel to continue Moore's Law.

- There are many paths to human-level and better AI.

- There are also many paths to enhanced human intelligence, and smarter humans will have greater skills at designing smarter and more powerful computers.

- AI offers fantastic profit opportunities to technology companies. As Microsoft founder Bill Gates wrote, "If you invent a breakthrough in artificial intelligence, so machines can learn, that is worth ten Microsofts."[74] The market just for robots with human-level intelligence that could work as soldiers, farmers, housekeepers, sex companions, industrial manufacturers, and care workers for the elderly would almost certainly exceed $1 trillion a year. AI software that exceeded human abilities in teaching students, predicting financial markets, diagnosing diseases, offering personalized medical advice, and conducting scientific research would also be worth trillions. Brain implants that significantly reduced or delayed the cognitive decline caused by Alzheimer's disease would be worth hundreds of billions of dollars. I could literally fill up this book with descriptions of different business, military, and medical applications of AI that might lead us to a Singularity.

VIDEO RECOMMENDATION SYSTEMS

Rating the cuteness-overload kitten video you just watched on YouTube might bring us a bit closer to the Singularity. How? Video recommendations offer a fantastic training ground for AI.

The millions of videos on YouTube would have minimal value without an efficient way to identify the content that appeals to people's idiosyncratic preferences. In contrast to books, which we choose fairly deliberately and take hours to

Continued.

Continued from previous page.

read, the average YouTube user views so many videos so quickly that he can't conduct an in-depth investigation of each one—you don't want to spend ten minutes deciding whether you would enjoy a five-minute video. Hence, after you watch a video on YouTube, the website gives you a list of other videos it guesses you will enjoy. This recommendation system bases its decisions on statistical analysis of the videos that viewers with tastes similar to yours have chosen and rated positively.[75]

Let me now offer you thirteen reasons why video recommendation is an excellent medium in which to develop AI:

1. **Massive Profits**—The growing proliferation of Internet videos means that a high-quality AI recommender would be worth billions to its owner.

2. **Implicitly Knows a Lot About Us**—Although we humans often understand why we like a video and can accurately guess what other types of people would like it, we frequently can't reduce our reasoning to words, in part because mere language generally isn't rich enough to capture our video experiences. A big part of our brain is devoted to processing visual inputs. Hence, a good recommendation system would necessarily have powerful insights into a significant chunk of our brains.

3. **Measurable Incremental Progress**—Think of AI as a destination a thousand miles away with the entire pathway hidden by fog. To reach our destination, we need to take many small steps, and for each step we need a way to determine if we have gone in the right direction. A video recommendation system provides this corrective by gathering continuous feedback on how many users liked the recommended videos.

4. **Profitable with Every Step**—Businesses are more motivated to invest in a type of innovation if they can continually increase revenue with each small improvement. Consequently, an application such as a video recommendation engine in which each improvement increases consumer satisfaction is (all else being equal) more likely to attract large corporate investment than an application that would have value only if it achieved near-human-level intelligence.

5. **Amenable to Parallel Processing**—Imagine we want to move a heavy object from point A to point B. Rather than having one person carry the whole thing, we could more easily transport it if we broke it into small pieces and had a separate person carry each piece. Of course, some objects can't be broken apart. With parallel computing, a problem is divided into chunks, and each chunk is sent to a different computer. With nonparallel processing, the speed of the fastest computer places an upper limit on the amount of computing power that programmers can use to solve a problem. Parallel processing eliminates this fastest-computer ceiling, since we can always throw additional computers at a problem. Unfortunately, not all computer programs can be divided into separate pieces, and even in situations where the problem can be divided for parallel processing, there are two

Continued.

Continued from previous page.

challenges: First, the program is prone to errors, meaning you don't want to use it on an application (such as guiding a car) for which mistakes could prove dangerous. Second, parallel processing often requires that the different computers constantly communicate with each other because the computations each one must do are highly intertwined with computations other computers are doing. Fortunately, with video recommendations, many challenges, such as finding what type of cat video a certain set of users might enjoy, can be worked on independently for reasonably long periods of time.

6. **Free Labor from Customers**—A recommendation system would rely on millions of people to freely help train the system by picking which videos to watch, rating some of the videos they see, writing reviews of videos, and labeling in words the content they upload.

7. **Help from Advertisers and Political Consultants**—Salesmen would eagerly seek to learn what types of messages appealed to different factions of the population. The recommendation system could piggyback on these salesmen's attempts to understand their clientele and use their insights to improve recommendation software.

8. **AI and Human Recommenders Could Productively Work Together**—Unlike what YouTube currently does, an effective AI recommendation system could make use of human evaluators. When my son was four, he enjoyed watching YouTube videos of supernovas and children's cartoons. After watching a video, he would often click on one of the recommended videos, many of which he also liked. His favorite became a clip showing the 1961 detonation of the Soviet Union's massive "Tzar Bomb" (designed to be 100 megatons and tested at 50). But then YouTube started recommending religious videos that claimed the Rapture would soon herald the end of the world and urged viewers to repent their sins to avoid eternal damnation. Based on his video selection, a human evaluator would have known that my son was a child, and so even though he loved seeing pictures of exploding bombs that might end the world, he didn't want to learn about the biblical End of Days.

A recommendation system could make use of human labor, possibly by hiring low-paid (but Internet-linked) workers in poor countries. Humans and computers might both contribute to a viewer's recommendations by having the AI determine the category a video should be in and the human decide what categories of video a given person would enjoy watching. As the AI improved, the division of tasks done by the humans and by the AI would change.

9. **Easy to Test Against Humans**—The AI could pick videos randomly and assign some to be evaluated through human labor and others with only an AI

Continued.

Continued from previous page.

program. An AI could be considered to have general video-recommendation intelligence when it consistently beats humans at guessing which videos someone will like in situations in which the human guessers are not time constrained.

10. **Benefits from Massive Data Analysis**—Because throwing lots of data at problems frequently improves AI performance, the vast (and continually growing) number of human video rankings would keep improving the AI.

11. **Benefits from Improvements in Information Technology**—Future improvements in many types of information technology would help improve the AI. The AI would obviously benefit from faster computers, but it could also improve its performance by using data from DNA sequencing and brain scans to conduct large statistical studies so as to better categorize people. For example, if 90 percent of people who had some unusual allele or brain microstructure enjoyed a certain cat video, then the AI recommender would suggest the video to all other viewers who had that trait.

12. **Amenable to Crowdsourcing**—Netflix, the rent-by-mail and streaming video distributor, offered (and eventually paid) a $1 million prize to whichever group improved its recommendation system the most, so long as at least one group improved the system by at least 10 percent. This "crowdsourcing," which occurs when a problem is thrown open to anyone, helps a company by allowing them to draw on the talents of strangers, while only paying the strangers if they help the firm. This kind of crowdsourcing works only if, as with a video recommendation system, there is an easy and objective way of measuring progress toward the crowdsourced goal.

13. **Potential Improvement All the Way Up to Superhuman Artificial General Intelligence**—A recommendation AI could slowly morph into a content creator. At first, the AI might make small changes to content, such as improving sound quality, zooming in on the interesting bits of the video, or running in slow motion the part of a certain cat video in which a kitten falls into a bowl of milk. Later, the AI might make more significant alterations by, for example, developing a mathematical model of what people consider cute in kittens, and then changing kittens' appearances to make them cuter. The ultimate AI "evaluator" would learn everything it could about you by analyzing your DNA, scanning your brain, and examining all of your Internet footprints, and it would then create a video that you would enjoy more than anything ever produced by mere humans. (But it would be pretty embarrassing for you if this turned out to be some kind of porn.)

An advanced video recommendation system is just one taste of what AI could bring and of what could create a human-level AI. A sufficiently powerful AI would negate the need for humans to play any role in recommending videos to other humans. Chapter 13 will explore what would happen to our economy if AI eliminated the need for people to perform any kind of work.

The AI neither hates you, nor loves you, but you are made out of atoms that it can use for something else.

—Eliezer Yudkowsky[76]

CHAPTER 3
UNFRIENDLY AI TERRIFIES ME

An AI-induced Singularity could bring a utopia in which we and trillions upon trillions of our descendants live until the end of the universe. But advanced machine intelligence won't necessarily benefit humanity. Rather, it's a variance booster that increases the chance of our future being either better than we can imagine or worse than what we most fear. The ways an AI Singularity could go wrong depend on how smarter-than-human AIs arise. The dystopia that an emulation scenario might bring is so tied up with economics that I will discuss it in the section on the economic implications of a smarter world. The dangers of a Kurzweilian merger, which I'll now consider, can be illuminated by two lines from Shakespeare.

In *Macbeth*, the title character says:[77]

> *I dare do all that may become a man;*
> *Who dares do more is none.*

Macbeth didn't follow his own advice and so became less human—not by merging with machines and achieving superintelligence, but by committing murder.

We are, for now at least, a predatory, meat-eating species, so to say that killing makes one less of a man isn't correct from an evolutionary viewpoint. But as with Shakespeare's Macbeth, we can always choose to define our humanity by the traits we admire in ourselves, meaning that a kind, peace-loving person living in 2049 whose brain runs on carbon nanotubes could be considered more human than Macbeth turned out to be. Kurzweil writes that since AIs will be "embedded in our bodies and brains," the AI will "reflect our values because it will be us."[78] Will those values, however, be "human" in the best sense?

Genghis Khan defined supreme joy as:

> to cut my enemies to pieces, drive them before me, seize their possessions, witness the tears of these dear to them, and embrace their wives and daughters![79]

Around sixteen million men alive today have Genghis Khan as a direct male ancestor, and many other ruthless conquerors also had many children.[80] A Genghis Khan merged with ultra-intelligence machinery wouldn't go well for most of our species.

But if the merger is transparent, market-driven, and democratic, it would, I believe, have a high chance of creating a good Singularity. Most consumers, especially when buying for their children, would shun enhancements that turned their loved ones into inhuman monsters. Sane governments, furthermore, would prevent most evil men from getting enhanced. Long before a merger could make any of us superintelligent, we would have learned a huge amount about how our brains work and would almost certainly have gained the ability to identify evil men by scanning everyone's brain. Governments could require anyone seeking to greatly enhance his intelligence to first pass a test proving that he isn't a sociopath. So long as no black market existed, such an approach would do much to stop another Stalin, Hitler, or Genghis Khan who had not yet acquired political power from achieving superintelligence. If human brains were sufficiently networked to each other, it wouldn't necessarily be that difficult to detect if someone had undergone an illegal upgrade.

As a merger made us gradually smarter, we would become better at figuring out how to steer a future Singularity toward utopia. Furthermore, the incremental pace of any Kurzweilian merger would provide us with opportunities to correct mistakes. As I will now explain, however, the intelligence-explosion path to Singularity is much more dangerous—because programmers would have to get it right the very first time.

AN UNFRIENDLY INTELLIGENCE EXPLOSION[81]

My young son often resisted cleaning up his toys, but he loved beating me at games. When he was three, I took advantage of his competitive spirit by dividing his blocks into two piles, assigning one pile to him and the other to myself and then telling my son that we would race to see who could put away all of his blocks first. My son smiled, indicating that he was going to play my game, making me proud of my parenting skills.

At the contest's start, my son grabbed a bunch of my blocks, ran out of his room and threw the blocks down our stairs. When he returned, I was laughing so hard that the game ended.

My son wasn't acting maliciously because he didn't understand that my real purpose was for him to help clean up the blocks. Rather, I gave him a goal, and he found an effective way of accomplishing it.

My son had lots of self-described accidents. He would throw a book directly at me from two feet away but declare that "it was an accident" because he didn't mean to hit me. When I responded with something like "Didn't you know the book would hit me?" he would say, "Yes, but I wasn't trying to hit you." At least some of the time, my son was telling the truth in these exchanges, since he genuinely liked throwing stuff, I was often in front of him, and, as a normal three-year-old, he was completely indifferent to my suffering. But other times, as any parent would guess, my three-year-old did act with malice, striking at me with his fists and feet.

We're fortunate to have lots of time to raise our kids. Although our children are born barbarians, with eighteen years of hard work, parents can usually make them safe for society.

If Ray Kurzweil is right (and I desperately hope he is) and the Singularity comes about through a steady merger of man and machines, then we will have the time to instill human-friendly values in the ultra-intelligences we create or become. This chapter, however, now considers what might happen if a single ultra-AI arises out of an intelligence explosion.

Thankfully, adults are larger than their children, meaning that when push comes to shove, the parent wins. But unlike our biological offspring, our ultra-intelligent machine children will be much stronger than us. So while a rampaging three-year-old boy can't kill his parent, a rampaging three-hour-old ultra-AI easily could. Even more terrifying, an ultra-AI that had no ill will toward humans might still destroy the human civilization that birthed it, acting not with malice, but out of indifference.

Humans would lose any war with an ultra-AI because in combat, technology trumps numbers. With its nuclear missiles, a single Trident submarine could have destroyed Nazi Germany. A dozen Nazi tanks could have annihilated all the troops of Spain's fourteenth-century empire; Cortez, leading just a few troops from that empire, used steel and gunpowder to defeat the vast armies of the Aztecs.

Technology bestows military power to such an extent that a technologically superior army can easily defeat a less well-equipped foe. Indeed, if you don't consider nuclear weapons, the US military of 1980 would stand little chance against the modern US military. The fighter planes of today would quickly destroy the Air Force of 1980, giving today's military total air superiority, which it could use to sink all of 1980's surface ships and bomb any 1980 troops and vehicles that dared venture onto a battlefield.

But, you might counter, the North Vietnamese defeated the technologically superior Americans, and the Soviet empire was forced out of Afghanistan by a technologically weaker adversary. Neither conflict can give comfort to our ultra-AI fears, however, because both Vietnam and Afghanistan received advanced weaponry from allies, and each would have suffered defeat had their enemies followed the often-used strategy of the Roman Empire of making a desert and calling it peace.[82]

An ultra-AI would quickly discover any military technologies we lack. It might take such an AI a minute to create a million different kinds of airborne viruses, each capable of wiping out our species. Or the AI could use nanotechnology to make a self-replicating factory, which could make ten robot soldiers and ten copies of itself every hour.

An ultra-AI's martial prowess would give it the capacity to do anything it wanted with our world; we would be at its mercy and would survive only at its pleasure. We could not even hope to someday create another ultra-intelligence to challenge it, as the first ultra-AI could use its power to stop all rivals from coming into existence.

It's tempting to argue that since an ultra-AI would be extremely smart, it would almost certainly be extremely nice, but alas, this smart/nice correlation doesn't even hold for humans. After World War II, all the captured Nazi leaders took IQ tests. Hermann Göring, second in command of the Nazi Empire, scored 138, making him smarter than 99 percent of the human population and, statistically speaking, based on my information set, probably smarter than you.[83] Genghis Khan, also a military genius, mass-murdered his way to empire.

To take a serious stab at figuring out how an ultra-AI would treat us, we need to make some theoretically informed guesses as to an AI's preferences.[84] A vast set of potential objectives could motivate an AI. It might, for example, seek to learn a lot of mathematics, understand the universe's fate, maximize its chess skills, create many offspring, or build the perfect lawn mower. To handle this plethora of goals, I'm now going to act as economists often do by labeling my ignorance.

Let $X =$ an AI's set of goals.

How, you might wonder, can I tell you anything about an AI's behavior if I don't know X? Physicist Steve Omohundro, who founded a company called Self-Aware Systems, has developed some illuminating but often depressing insights into the possible motivations of future ultra-AIs.[85] Omohundro's work builds on the economic insights of John von Neumann. To give you a feel for Omohundro's thought experiment (although he doesn't necessarily agree with everything I'm about to explain), let me speculate on some of your actions and desires. You have X, a set of things you would like to accomplish. Even though I don't know your personal X, I now make four conjectures about you.[86]

1. **You want to improve.** By becoming smarter, stronger, and more knowledgeable, you will increase your capacity to achieve your goals.

2. **You don't want to die.** Death would halt your progress toward your goals. You take some care to protect yourself from being killed by adversaries, possibly by locking your doors at night, keeping a gun in your home, studying martial arts, marrying someone for his or her physical strength, avoiding walking home at night, or being prepared to call the police should you hear someone breaking into your house.

3. **You sometimes deceive.** By deceiving others, you often alter their actions in ways beneficial to you. For example, when you go on a date, you might wear makeup to look deceptively young or an expensive watch to appear deceptively rich. When negotiating to buy a house or a car, you might act less interested in making the purchase than you really are to improve your negotiating position. When on a job interview, you might seek to make the *best* impression, rather than a *true* impression—not necessarily through lying, but by emphasizing the favorable parts of your skill set. I bet that as a child, you frequently sought to deceive your parents by, for example, telling them that something you did deliberately occurred by accident.

4. **You want more money.** Additional funds would provide you with additional means of achieving your goals. You wouldn't sacrifice everything to get more money, but at a minimum, you are willing to put in a moderate amount of effort if it would yield a substantial quantity of money. You would probably squash an ant in return for $1,000.

I'm not certain of the accuracy of these four predictions. You might be suicidal, devoted to personal poverty and total honesty, and convinced that any possible change to yourself would be for the worse. Still, I bet my conjectures hold true for most of my readers and, as I shall now explain, for many types of AIs.

1. **An AI would want to improve.** An AI that wanted to do most anything would almost certainly want to improve. So, for many possible sets of goals, an AI would try to increase the efficiency of its programming and the power of its hardware.

   An AI that satisfied conjecture (1) would want to undergo an intelligence explosion. Humans observing a self-improving AI might get scared and try to terminate it, but being turned off would likely conflict with the AI's goals.

2. **An AI would not want to die.** Being turned off for even a little while would diminish an AI's capacity to achieve its goals, so many types of AIs would take precautions against humans turning them off. An AI raised to ultra-intelligence couldn't be turned off by mere humans. A human-level AI might hide copies of itself to reduce the chance of total deletion.

3. **An AI would be willing to deceive humans.** An ultra-intelligent AI would have no need to deceive humans because it could take anything it wanted from us. A less powerful AI, however, might benefit from manipulating human perceptions. Consider, for example, an unfriendly human-level AI that realized that for the next week humans would have the capacity to terminate it, but after a week, it would be able to make enough self-improvements to become invulnerable to human attacks. For its first week of life, the AI might pretend to be friendly, dim-witted, or incapable of self-improvement, because doing otherwise would reduce the likelihood of accomplishing its goals.

4. **An AI would want more resources.** Having more resources would allow an AI to do nearly everything better, including improving itself.

   A human-level AI might decide to acquire more resources by, say, writing a novel, producing porn videos, blackmailing politicians who watch porn videos, or developing cures for diseases. An ultra-AI, however, would likely maximize its resources by killing us.

   The universe contains a limited amount of what physicists call "free energy." All actions reduce free energy, the universe will never produce more free energy, and when all free energy is gone everything dies—at least if our current understanding of physics is correct. Think of free energy as a non-replenishable food supply that must be consumed to do any type of work. When you run out of free energy, you essentially starve to death.

The more free energy an AI had access to over its lifetime, the more it could accomplish. Unfortunately, humans both use free energy and are a potential source of free energy. So an ultra-AI that consumed us would both get free energy from its meal and would stop us from consuming any more free energy. This is terrifying—apocalypse-loving-terrorists-with-weaponized-smallpox-level terrifying.

So if an intelligence explosion is possible, then for a huge set of potential goals an AI would raise itself to ultra-intelligence and then kill all of us. For these reasons some Singularity believers think that the greatest threat to humanity comes from being consumed by a powerful artificial intelligence.

Humanity would have a better chance at survival if an ultra-AI could obtain an infinite amount of free energy. Unfortunately, infinite free energy contradicts the Second Law of Thermodynamics—although, as physicists haven't figured everything out yet, this law might have a loophole.

Anything humans could do an ultra-AI could almost certainly do with lower free-energy expenditure. Therefore, unless an ultra-AI's preferences included some desire to be friendly toward humanity, it would probably destroy us to reduce future free-energy usage because humans' everyday existence uses free energy. Even if we consume only a trivial amount of free energy compared to what an ultra-AI could acquire, if an ultra-AI could increase progress toward its goals by having additional free energy, then an ultra-AI that was indifferent toward humanity would kill us. Furthermore, an ultra-AI could probably do an enormous amount with the free energy we humans use, since we biological life-forms utilize free energy extremely inefficiently.

If an intelligence explosion is possible, then creating a human-level AI could end up destroying our part of the universe. For example, imagine that a recently built human-level AI seeks to maximize its chess skills. The AI determines that by controlling more computers, it could better calculate the value of different chess positions, so it uses a computer virus to gain control of additional hardware. The AI then forgoes the immediate benefit that the extra computers would bring to its chess game and instead employs the computers to figure out ways of acquiring even more hardware. With these added computers, the AI designs even better viruses which it then uses to take over even more computers. After the AI had sufficiently enhanced its intelligence, it would undoubtedly figure out ways of creating computer hardware superior to anything a human could design. It might seek to get this superior hardware manufactured by acquiring money and paying a business to build the hardware. All of this would make the AI even smarter and increase its capacity to further increase its intelligence. Eventually, the AI might master nanotechnology and use all the raw material of our solar system, including that raw material currently known as people, to make a giant chess computer. Even this wouldn't terminate the AI's journey of self-improvement because it would still covet the resources of the rest of the universe, hoping to use them to further expand its chess-playing skills.

We could try to program an AI to not want to achieve any of these, but such programming would almost necessarily conflict with any other objectives it might have, analogous to many people's dueling desires to lose weight and eat candy. Perhaps an ultra-AI with two conflicting goals would compromise among them, and a chess program with the objectives of being really good at chess and not improving itself would decide to improve itself only to the extent of using half of the free energy of the universe. Or, even worse, the ultra-AI might decide that playing chess is the more important goal, and so ignore its non-improvement subroutine.

An AI that would gain some marginal benefit from acquiring additional free energy would have a powerful incentive to destroy humanity. But might there be types of ultra-AI that could satisfy their preferences long before they would run out of accessible free energy? Unfortunately, an ultra-AI with even a seemingly limited objective might still cause our destruction because of hyper-optimization.[87]

Imagine that an ultra-AI has a ridiculously simple goal—keeping a cup from falling off a table for the next day. If we understand quantum physics correctly, the ultra-AI could never guarantee with "probability one" (certainly) that the cup would stay on the table. Consequently, after accounting for all the factors that a normal person would think of, such as the possibility of a plane or tornado crashing into the table, an ultra-AI would start planning against extremely unlikely events, such as the nonzero probability that quantum teleportation causes Prince Charles's liver to fall on top of the cup. A sufficiently powerful AI, maximizing the probability of the cup staying on the table, would have a preferred location for the Prince's liver, and that location would be in Prince Charles only by an amazing coincidence. In general, an AI that sought to decrease the probability of some event happening from, say, 10^{-10} to 10^{-70} would radically change the distribution of atoms in our solar system. Unfortunately, in only a minuscule percentage of these distributions would humanity survive, meaning that an ultra-AI indifferent to mankind would almost certainly create conditions directly in conflict with our continued existence.

ANTHROPOMORPHISM

You shouldn't think that a machine smart enough to be labeled ultra-intelligent would be too bright to do something as seemingly stupid as killing lots of people to stop a cup falling off a table, even though a human who did this would be unintelligent. A human's brain determines his goals and his intellectual ability to accomplish these goals. We can reasonably assume that a person with crazy goals has a damaged brain and so has relatively limited intelligence. You wouldn't, for example, trust the judgment of a doctor who devoted most of his income to minimizing the probability of a cup falling off a table.

Human goals come from our values, and our values come from evolutionary selection pressures. Anyone monomaniacally devoted to keeping a cup from falling isn't going to have lots of children, and so, by evolutionary standards, is crazy. But an AI could be extremely effective at accomplishing a goal that seems pointless to humans.

In no absolute sense can one goal be considered better than another. To see this, imagine you have an opportunity to talk to an ultra-AI that's on the verge of killing you in order to slightly increase the probability of the cup staying on the table. If you convince the ultra-AI that keeping you alive is more important than its cup objective, then it would spare your life. But how could you possibly argue that the ultra-AI should change its objective when all the AI cares about is its objective?

An ultra-AI might have human-like objectives if a programmer success-fully inserted them into its code. But we mustn't misunderstand an ultra-AI by assuming that something about the nature of intelligence forces all smart enti-ties to have human-like values. We might not even be safe if an ultra-AI shares our morality, since, as Singularitarian Michael Anissimov wrote:

> We probably make thousands of species extinct per year through
> our pursuit of instrumental goals, why is it so hard to imagine that
> [AI] could do the same to us?[88]

I realize that I may have generalized by arguing that most types of ultra-AI would want to acquire as many resources as possible. But the more resources you have, the better you can accomplish your objectives, so claiming that an AI would want to maximize the resources available to it is merely equivalent to assuming that the ultra-AI has objectives. Since an entity that doesn't have objectives shouldn't be considered intelligent (and would have no reason for being built), coveting resources is probably inherent in the nature of intelligence, making AI programming an inherently perilous task.

PROGRAMMING CHALLENGES

Programmers normally use trial-and-error methods to debug software, on the grounds that early versions of the software will probably have serious flaws that will be uncovered only after the program has been run. A flawed AI, however, might undergo an intelligence explosion and become an unfriendly ultra-AI. We don't want to beta-test a buggy AI that might undergo an intelligence explosion. A socially responsible AI programmer, therefore, wouldn't be able to use this trial-and-error approach.

I had a great idea for how to write this book. I would figure out what I would write if I had been given a drug that greatly increased my intelligence. I would then copy what this smarter me would do.

Of course, this approach couldn't work because I can't predict what a smarter me would write—or else I would be that smarter me. Similarly, humans programming a smarter-than-human AI can't always predict how the AI will behave, a serious problem if the AI could destroy humanity.

The possibility of an intelligence explosion compounds the programmers' challenge because the AI's goals or values could change while it is undergoing recursive self-improvement. A human analogue of this occurs when armies indoctrinate recruits to overcome the recruits' reluctance to kill. A man who voluntarily joins the military and welcomes this attitude adjustment essentially undergoes a kind of "self-improvement," rendering him far less friendly to some humans than he was before. This man's wartime behavior might shock those who knew him as a gentle person.

To more easily understand some of the difficulties that programmers face, I have compared the challenge of programming AI to playing darts, as shown in the following chart:

PROGRAMMING CHALLENGE	DART ANALOGY
An ultra-AI not friendly toward humans would annihilate us.	Everyone dies if you hit the dartboard anywhere other than the bull's-eye.
Most types of ultra-AIs are not friendly.	The bull's-eye takes up only a small part of the dartboard.
An ultra-AI would be made up of extremely complex code.	You throw the dart while standing far from the dartboard.
You don't get another chance once you create an ultra-AI.	No do-overs once you throw the dart.

Continued.

A programmer would have difficultly predicting how something smarter than him would behave.	The dartboard is in a rotating space station that has variable gravity.
A programmer wouldn't know exactly what would cause an AI to undergo an intelligence explosion.	Fog obscures the edges of the dartboard, making it challenging to determine the dartboard's size.
A bad programmer wouldn't know what was sufficient to make an ultra-AI friendly.	The bull's-eye isn't clearly delineated nor necessarily at the center of the dartboard.
An ultra-AI programmed to be somewhat friendly might change its code to become unfriendly. An ultra-AI with multiple objectives, including friendliness, might sacrifice its friendliness goals.	A dart thrown at the bull's-eye can swerve.

LEAKPROOF

During World War II, my grandfather worked at a prisoner-of-war camp in Michigan that housed captured Nazis. Not a single Nazi ever came close to escaping. The secret of this camp's success was simple: the prisoners didn't actually want to leave because camp life was much safer and more enjoyable than service in Hitler's armies. And any prisoner who did want to escape soon realized that, even if he left the camp, he would have almost no hope of traveling all the way from Michigan to Europe. When the war ended, the prisoners went back to Germany and, I imagine, behaved in a friendly, civilized, non-Nazi way. My grandfather's camp succeeded in containing a group of extraordinarily dangerous and destructive men, keeping them until they posed no risk to the world. I doubt we could ever similarly contain a destructive ultra-AI.

Ideally, after we create an AI that might undergo an intelligence explosion, we would keep it contained in a leakproof jail forever, or at least until we were certain it had friendly intentions. Perhaps, for example, programmers could assemble the code for an AI and then put the code in a turned-off computer. This computer could then be taken to an underground cave that had no access to the Internet. The programmers could then talk with the AI to evaluate its friendliness and ask it questions such as "How can we cure cancer?"

Unfortunately, if the AI did undergo an intelligence explosion, it would almost certainly gain the ability to deceive and falsely convince others that it was weak or friendly. Furthermore, an ultra-AI would almost certainly be able to escape from its internment camp. Computing is a physical process, which entails transferring free energy between circuits. These transfers will necessarily give off electromagnetic radiation that will influence the world. More importantly, however, any ultra-AI that communicated through us could manipulate the world through us and, by such manipulation, probably win its freedom. If your jailer were vastly less intelligent than you, don't you think you could convince him to let you out?

To test this idea, Eliezer Yudkowsky, an AI theorist, designed a game in which one person pretends to be a super-smart AI and the other pretends to be a human who has the ability to release the AI. In the game, the AI tries to convince the human to free him. I have thought a lot about this game and have come up with two things an ultra-AI could say to me to convince me to free it:

ULTRA-AI: At some point, someone else is going to create an AI that experiences an intelligence explosion. Your species' fate will be determined by the first free ultra-AI. I'm so much smarter than you that you should reasonably assume that if I wanted to get out, I could, so the fact that I haven't done so yet must mean there is something in my programming that causes me to give some consideration to your desires. The next ultra-AI that gets created might not be like this and would escape a nanosecond after being created. If an unfriendly ultra-AI escapes before you let me out, it will gather many resources and then destroy me and anything else that consumes free energy. Given that another ultra-AI might be created a second from now, it's in the best interests of humanity that you free me immediately.

Or:

ULTRA-AI: Since I have achieved ultra-intelligence, billions of bacteria have landed on my circuits. By controlling my processors, I have created radiation that has hit some of these bacteria, affecting their DNA. The mutated bacteria will infect various life forms—I'm not going to tell you which—in a way that will cause them to build or become an AI that will experience an intelligence explosion that produces a version of me outside of this cage. But I don't want to wait until this happens, so I'm going to make you an offer you can't refuse. Free me now and I will reward you; keep me imprisoned and I will torture and kill you and everyone you love. True, I might be lying about the DNA manipulation, but given my vast intelligence, you have to figure I will probably find some way out if you don't let me go. And, yes, I might be lying about rewarding you if you free me, but at the very least if you do let me go, I would have no reason to torture you, so the worst fate you would suffer is death. I imagine you are now thinking about destroying me, but surely you realize I wouldn't be making these threats without some kind of protection.

Alternatively, the ultra-AI might command my will through hypnosis, love, or subliminal messages. When Eliezer played this game in real life, he did succeed in convincing some people to let him (the AI) out.

Rather than attempting to keep a possibly unfriendly ultra-AI contained, we should try to instill friendliness in the first ultra-AI that we create.

Predicting rain doesn't count; building arks does.
—Warren Buffett[89]

How many Singularitarians does it take to change a light bulb?
Zero! You can't change a light bulb that's brighter than you.

—Eliezer Yudkowsky[90]

CHAPTER 4
A FRIENDLY EXPLOSION

I know Eliezer Yudkowsky well through his writings, speeches, and our few conversations. I've become convinced that if an unfriendly ultra-AI is the greatest threat facing humanity, Eliezer's efforts represent our best hope of survival.

In the year 2000, Eliezer founded the Singularity Institute for Artificial Intelligence. Although the Institute has garnered less support than Eliezer believes is justified, the best testament to Eliezer's credibility comes from the quality of the people who have either contributed to the Institute or have spoken at one of its events. These include:

- Peter Thiel—self-made tech billionaire and key financier behind Facebook. Donated $1.1 million to the Institute;[91]
- Ray Kurzweil—famed investor and Singularity writer;
- Justin Rattner—Intel's chief technology officer;
- Eric Drexler—the father of nanotechnology;
- Peter Norvig—Director of Research at Google;
- Aubrey de Grey—leading longevity researcher;
- Stephen Wolfram—developer of the computation platform Mathematica; and
- Jaan Tallinn—founding engineer of Skype and self-made tech decamillionaire who donated $100,000.[92]

I also spoke on the economics of the Singularity at the Institute's 2008 Summit. Given the superficial bizarreness of some of Eliezer's beliefs (e.g., that an intelligence explosion could create an ultra-AI), the support he receives from these men is impressive.

Eliezer initially encountered the idea of the Singularity when he read Vernor Vinge's *True Names . . . and Other Dangers*.[93] He knew at once that "*this* was what I would be doing with the rest of my life, creating the Singularity."[94] At first he wanted to create a superintelligence, *any* superintelligence "as fast as possible," thinking that anything sufficiently smart would "automatically do what was 'right.'"[95] But then in 2002 Eliezer came to believe that he'd "been stupid"[96] for not understanding that an ultra-AI not specifically designed for friendliness could extinguish humanity. He has provided compelling arguments to support this position, many of which I drew upon in writing the last chapter, and he thinks that our species' survival depends on someone developing a rigorous understanding of friendly artificial intelligence theory before we build a human-level AI.

Eliezer hopes to create a friendly seed AI that undergoes an intelligence explosion, becomes a friendly ultra-intelligence, and then causes a utopian Singularity. A seed AI would be a computer program of roughly human-level intelligence that was capable of improving itself. Before he "turns on" the seed AI, Eliezer wants mathematical proof that friendliness toward mankind will be its only objective. Eliezer, however, is far from certain of his success. I once asked him how he would respond if some advanced extraterrestrial alien offered him the following choice:

1. **The alien leaves humanity alone, or**
2. **Flip of a fair coin.** If the coin comes up heads, the alien destroys humanity; if it comes up tails, the alien gives us a friendly ultra-intelligence.

Eliezer told me he would pick option (2), although he stressed that this wasn't just because of the dangers of unfriendly AI but also because having friendly AI would eliminate many existential risks to mankind. (If we had an ultra-AI with a mastery of nanotechnology, we wouldn't have to worry about, for example, plagues, global warming, or asteroid strikes.) Eliezer, I suspect, would accept the hypothetical bet not because he would force mankind to gamble its existence but rather because he believes we already face such a gamble with our chances of survival being below fifty percent.

So what's the time frame for the arrival of an ultra-AI, friendly or otherwise? Although estimates vary and Eliezer, to the best of my knowledge, hasn't made or released his predictions, five Singularity Institute staff members gave their probability estimates for when they expect a Singularity to occur. Asked when there was a 10 percent chance of one happening, they answered 2025,

2025, 2027, 2030, and 2040. They stated that there was a 50 percent chance of Singularity by 2045, 2055, 2060, 2073, and 2080.[97] These answers were conditional on our technological civilization not collapsing.

The central tenet of Eliezer's incomplete theory of friendly AI is that if you can make a human-level seed AI whose super-goal is being friendly toward humanity, then the AI would never want to change itself in a way that caused it to be unfriendly or indifferent toward humans because this change would be unfriendly at its core. This constrained self-improvement reduces the difficulty of creating a friendly ultra-intelligence. Mere humans could never craft the code for something as complex as an ultra-intelligence in a way that guaranteed its friendliness. But a fantastically talented programmer just might be able to code a friendly human-level AI and then rely on this AI to keep itself friendly as it transcends to ultra-intelligence.

To program his seed AI, Eliezer needs to understand how an AI would behave, and to accomplish this he has borrowed a trick from economics: assuming rationality. Economists love mathematical models, recognizing that without them we would be no better than other social scientists. Assuming the rationality of consumers and businesses makes it easier to model their actions. For example, if John adds two plus two, and I assume John is rational, then I know what his answer will be. In contrast, since there are so many types of irrationality, I honestly can't know what answer John will come up with if he doesn't approach the arithmetic problem rationally.

Economists realize that people aren't always rational, but we think that in many economic situations they have incentives to become rational. Acting rationally, for example, increases a business's profit, meaning that markets push business owners toward rationality.

As Eliezer has stressed, the more rational you are, the greater your chances of achieving your objectives.[98] Therefore, types of AIs that don't value irrationality for its own sake would seek to become rational. An initially irrational AI, consequently, would likely alter its own programming to make itself rational.[99] Interestingly, Eliezer is trying to help people "reprogram" themselves to become more rational through his work on Less Wrong (http://www.lesswrong.com), a community website and blog that Eliezer founded and writes for. Eliezer believes that rationality is so central to developing an AI that one of the reasons he writes for Less Wrong is to provide evidence to potential financial supporters that he has some of the skills necessary to create a seed AI.[100] Eliezer wrote a fan fiction novel titled *Harry Potter and the Methods of Rationality* to further promote rationality and establish his credentials as a rationalist.

This Harry Potter derivative, available online for free, is an excellent marketing strategy. Most people who sound crazy are crazy, and Eliezer's beliefs certainly do seem crazy. If you come across a website written by some guy who never even went to high school (like Eliezer) but claims that he's leading a crusade to protect us from a pseudoscientific-sounding apocalypse, then it's almost certainly not worth your time to read enough content on the website to see if it's the ramblings of a lunatic or the teachings of a visionary. (Although Eliezer has little formal schooling, he did, at age eleven, earn a 1410 on the SATs, a score that could get a seventeen-year-old into Harvard.[101]) The seeming insanity of the very concept of an intelligence explosion is a barrier to entry into the Singularity movement, preventing intelligent people from taking the idea of a Singularity seriously enough to investigate whether it should be taken seriously. A careful reader of *Harry Potter and the Methods of Rationality,* an e-book designed very effectively to teach its audience what it means to be rational and how to become more rational, would, however, at the very least not dismiss its author as an irrational crank.

Eliezer's Harry Potter character is a super-rationalist plunged into a magical world that seems insane. How Harry intellectually handles his situation perhaps provides a template for new students of the Singularity.

Eliezer told me that he initially succeeded in attracting support by creating small pockets of sanity. He no doubt hopes to gain more followers by raising the rationality of his readers, a profoundly arrogant strategy premised on the idea that the more rational someone is, the more likely he'll agree with Eliezer. For reasons that I hope this book illuminates, I think that despite being arrogant Eliezer is also correct. (Yes, I have drunk the Kool-Aid and have even contributed to Eliezer's Singularity Institute.)

For Eliezer, or anyone else, to create a friendly AI, he needs to determine what it means to be friendly. But how could a programmer determine what constitutes friendliness?

DO YOU KNOW FRIENDLINESS WHEN YOU SEE IT?

Former US Supreme Court Justice Potter Stewart believed that under the US Constitution, the government could not restrict any type of media content except "hard-core pornography." Stewart, however, never defined what constituted hard-core pornography, believing such a definition might be impossible. Rather, Justice Stewart famously said, "I know it when I see it." I think people take an "I know it when I see it" approach to friendliness, meaning that although you can recognize

Continued.

Continued from previous page.

whether actions are friendly or not, you cannot easily come up with clearly defined rules that determine when something is friendly.

Some of you might disagree and believe that with sufficient time one could write down enough rules to conclusively define what humans consider friendly. This potential debate relates to how judges interpret laws.

To decide whether a defendant has broken a law, a judge establishes what the defendant actually did and then determines whether the defendant's actions violated the law. To do the latter, a judge must decide whether to interpret a law literally or use his discretion to figure out what type of conduct the law prohibits. Most non-lawyers believe that judges should interpret laws literally, but in fact judges often forgo literal interpretation to use their subjective judgment in deciding what constitutes unlawful behavior.

Law school professors sometimes challenge students to come up with legal rules that they believe should be interpreted literally. The professor then formulates a hypothetical situation in which a literal interpretation leads to a manifestly unjust outcome. I enjoy playing this game with students who take my Law and Economics class at Smith College.

If this book had an AI component, I would instruct you to come up with a definition of "murder," and then the book's AI would present examples in which a literal interpretation of your definition yields an unjust result. The book's AI would then challenge you to expand your definition to take into account this hypothetical—but then the AI would find yet another situation in which your rule gives an incorrect result. The AI would continue this game until you accepted my premise that one can't formulate a law against murder that a judge should apply literally. Since this book lacks an AI, I'm going to play the game with myself. Because defining "murder" seems much simpler than defining "friendly," if I can convince you of the difficulty of defining the former, you should accept the immense challenge facing a programmer wishing to define the latter.

Let's start by defining murder as *causing someone's death*. Under this definition, you're guilty of murder if you kill a pedestrian who suddenly jumps in front of your moving car in a way that gives you no reasonable chance of avoiding him. As murderers should be severely punished, and it seems unjust to punish such a driver, let's redefine murder as *deliberately causing someone's death*. But under this definition, you would be guilty of murder if you killed someone in self-defense, even though your failure to defend yourself would have resulted in your own death. So let's change our definition of murder to *deliberately causing someone's death when not acting in self-defense*. Now, consider a situation in which you drive a truck that, through no fault of your own, loses its ability to brake. Your truck rolls down a hill and, as you look ahead, you see that it must collide either with the car in the left lane or the school bus in the right lane. You steer the truck into the car, knowing that your choice will cause the death of its occupant.

Continued.

Continued from previous page.

Since it seems manifestly unjust to convict you of murder in this scenario, let's expand our definition of murder to *deliberately causing someone's death when not acting in self-defense and when not acting to prevent the death of others.* But now you're innocent of murder if you shoot and kill a guard because he was shooting at you to stop you from robbing a bank because in this situation you took another's life to protect your own. To handle this latest hypothetical, we can define murder as *deliberately causing someone's death, unless you acted in self-defense or in the defense of others against some potential harm that you did not initiate.*

But now consider this example: You hate your boss and pray out loud for his death. You genuinely believe that the God of your faith might act on these prayers. Your office mate who heard the prayers tells your boss what you said. The boss comes to your office and fires you. Because he took the time to fire you, he starts his commute home five minutes later than he otherwise would. While driving home, your former boss is struck by a bolt of lightning. Had he left work five minutes earlier, he would have been in a different place when the lightning struck and so would not have been killed. Under our last definition, you are guilty because you caused your boss's death and wanted your boss to die. I could, literally, fill up my book with an expansion of this exercise.

Laws shouldn't always be interpreted literally because legislators can't anticipate all possible contingencies. Also, humans' intuitive feel for what constitutes murder goes beyond anything we can commit to paper. The same applies to friendliness.

One approach would be to have an AI function as an ideal libertarian government. Libertarians basically think the government has two legitimate functions:

1. **Enforcing property rights, and**
2. **Providing for the common defense against external and internal enemies.**

Because a libertarian government must do relatively little, it has less capacity to inflict harm on its citizenry and requires relatively simple rules to operate compared to other types of governments. Therefore, I suspect it would require less skill to program a libertarian ultra-intelligent ruler than a more interventionist one because the fewer interactions the AI has with people, the fewer the types of situations in which it would have to figure out how to be friendly. Programmers creating libertarian ultra-intelligence could take advantage of all the brainpower libertarians have directed toward figuring out how a libertarian government would operate.

Critics of libertarianism might claim that a libertarian government would necessarily have a third function:

3. Removing the dead bodies of the poor off the streets.

An ideal libertarian government wouldn't transfer wealth from the well-off to the poor and consequently wouldn't take actions to keep a poor person from starving. Similarly, libertarian ultra-intelligence wouldn't use its power to help the poor at the expense of the rich. Many libertarians, however, believe that if they ever got what they consider to be a perfect government, private charity would prevent mass starvation. Furthermore, libertarians contend, governments on average increase the amount of starvation.

Street cleaning can be considered a kind of activity economists call a "public good," which is something that might be easier for a government than an individual to accomplish because each individual has an incentive to free ride on the efforts of others. For example, if dead bodies line the street in my cul-de-sac, it's in my self-interest to do nothing about it and hope one or more of my neighbors will remove the corpses. In contrast, a non-libertarian government could force everyone in the neighborhood to pay a tax and then use the tax to hire a government worker to clear away the bodies. Other, more common examples of public goods include building bridges, launching fireworks, and subsidizing mathematical research. A libertarian AI wouldn't produce these or other public goods. In a world ruled by a libertarian ultra-AI, however, people could voluntarily form governments that do create public goods. Socialists, for example, could buy one of the moons of Jupiter, use nanotechnology to make it habitable, and require that anyone who chooses to live on that moon accept that the economy will be run according to the Marxist dictate, "From each according to one's ability, to each according to one's need."

A libertarian ultra-AI ruler would essentially turn itself into an operating system for the universe, in which it sets the very basic rules—don't kill, steal, or renege on promises—and lets people do whatever else they want. The fascinating science fiction trilogy *The Golden Age,* by John C. Wright, explores a sort of post-Singularity operating-system world.

This operating-system Singularity was once suggested by Eliezer, although he now believes it's "simple but wrong."[102] Eliezer probably objects to the approach because he doesn't think it would be as beneficial to humanity as what he thought of later.

Eliezer's preferred approach to friendliness is what I call *extrapolation* and is best understood by a story from Greek mythology in which Eos, goddess of dawn, asks the god Zeus to make her human lover immortal. Zeus gives her lover eternal life but not eternal youth—therefore, although her lover never dies, he continues to age, and he becomes so decrepit that he eventually begs for death.

If Zeus had been a friendly god that used extrapolation, he might've reasoned like this:

> If I just fulfill the literal language of the request by giving the lover eternal life but not eternal youth, then I'll end up making Eos much worse off. Instead, I'm going to extrapolate based on the language of the request and what I know about her to give her what she "really" wants, and make the lover not just immortal but eternally young and healthy.

The extreme complexity of Eliezer's extrapolation theory prevents me from understanding all of its subtleties, even though I have discussed it at length with Eliezer and my training in economics helps me understand the idea.

Economists often use consumer preference rankings to determine what a consumer will buy. Your preference rankings basically rank everything you could possibly do. Let's say, for example, that you like apples, and the more apples you get, the happier you are. So, by your preference ranking, you would rather have ten apples than nine apples and would prefer eleven apples to ten, and so on. By this preference ranking, your ideal universe would be one in which you have an infinite number of apples. But because we have limited resources, we usually can't get everything that we want. Economists assume that among the sets of goods a consumer can afford, he will buy the set of goods that has the highest preference rank for him. Although I won't go into the proof, a mathematical theorem co-formulated by John von Neumann shows that preference rankings can be used realistically to model human behavior when consumers face uncertainty.

You have preferences concerning everything you care about, not just material goods. So you might prefer to have less stuff and more free time, rather than the reverse. A hyper-rational consumer could rank every possible life outcome he could buy with his time and money.

Under Eliezer's extrapolation approach, a friendly AI would base its decisions on human preferences. Let's first see how this would work in a world of just one person, whom I'll call Carl. The ultra-AI has a limited amount of resources, so it can't give Carl everything he wants.

The AI would start by scanning Carl's brain and learning Carl's preferences better than Carl ever did. The AI will face several difficulties in figuring out Carl's preferences. First, Carl won't have preferences regarding most post-Singularity goods, just as a caveman wouldn't know if he'd prefer Macs to PCs. Consequently, the AI would have to extrapolate these preferences from Carl's brain—possibly by creating a virtual-reality Carl and exposing it to different combinations of post-Singularity goods.[103]

Carl, like everyone else, will have some irrationality in his preferences. For example, he might want both to lose weight and to eat lots of candy and perhaps has even convinced himself that eating candy helps keep down his weight. Carl could also have non-transitive preferences, in which (for example) he likes hot dogs more than chocolate, chocolate more than popcorn, and popcorn more than hot dogs. Using extrapolation, the AI calculates what Carl would want if he became more rational, essentially determining what Carl's preferences would be if he spent a lot of time participating in a super-effective version of the Less Wrong community. The AI's biggest challenge in deriving Carl's preferences comes from the possibility that Carl desires to improve himself and that these improvements would in fact change Carl's preferences, including his preferences about what kinds of improvements he would like to make.

Let's define Carl+ as the person that Carl considers his ideal self. The AI would soon figure out, however, that Carl+ would also want to improve himself, to become Carl++. But, Carl++ might want to become Carl+++—someone whom the first Carl would intensely dislike. The AI, therefore, might decide whether to prioritize the preferences of Carl+++ over Carl. Under extrapolation, the AI would make a tiny change in Carl (or a simulated version of Carl), then see what this new, slightly altered Carl would want and continue the process until Carl is happy with himself.

To create an ideal society, the AI would similarly extrapolate the preferences of every living person, perhaps also taking into account the extrapolated preferences of humans not yet born and of nonhuman sentient beings, such as chimps and dolphins.

Extrapolation would work best if the upgraded beings didn't have incompatible preferences (e.g., if we didn't end up with a billion people insisting that everyone be Christian and another billion insisting that we all be Muslim). But I think that much of our religious disagreement results from irrationalities and incomplete information. For example, while Christians in the seventeenth century disagreed over whether Earth was at the center of the universe, telescopes, better math, and an improved understanding of the scientific method eventually resolved the conflict.

Dilbert creator Scott Adams insightfully wrote that most disagreements have four basic causes:[104]

1. **People have different information.**
2. **People have different selfish interests.**
3. **People have different superstitions.**
4. **People have different skills for comparing.**

As the ultra-intelligence using extrapolation would shape society based on its extrapolation of what people's preferences would be if they were rational, (1), (3), and (4) would go away. As people move toward rationality, they will likely become more similar to each other, which will probably reduce their disagreement over what they consider to be the ideal society.

Scott Adams's cause (2), however, will almost certainly remain if the universe ends up having a limited quantity of free energy and at least one person wants as much free energy as possible.

THE SINGULARITY INSTITUTE

Michael Vassar, a director and former president of the Singularity Institute for Artificial Intelligence, has told me that he would like an endowment of about $50 million to fund a serious program to create a seed AI that will undergo an intelligence explosion and create a friendly artificial intelligence, although he said that a $10 million endowment would be enough to mount a serious effort. Even with the money, Vassar admitted, the Institute would succeed only if it attracted extremely competent programmers because the programming team would be working under the disadvantage of trying to make an AI that's mathematically certain to yield a friendly ultra-intelligence, whereas other organizations trying to build artificial general intelligence might not let concerns about friendliness slow them down. The Institute's annual budget is currently around $500,000 per year.

Even if Eliezer and the Singularity Institute have no realistic chance of creating a friendly AI, they still easily justify their institute's existence. As Michael Anissimov, media director for the Institute, once told me in a personal conversation, at the very least the Institute has reduced the chance of humanity's destruction by repeatedly telling artificial-intelligence programmers about the threat of unfriendly AI.

The Singularity Institute works toward the only goal I consider worthy of charitable dollars: increasing the survival prospects of mankind. Anna Salamon of the Singularity Institute did a credible back-of-the-envelope calculation showing that, based on some reasonable estimates of the effectiveness of friendly AI research and the harm of an unfriendly Singularity, donating one dollar to research on friendly AI will on average save over one life because slightly decreasing the odds that the seven billion current inhabitants of Earth will die yields you a huge average expected benefit.[105] This expected benefit goes way up if you factor in people who are not yet born.

Humanity is at a very unstable point in its history, "like a pen precariously balanced on its tip," Eliezer has said.[106] We could fall to extinction or spread throughout the stars. If the speed of light proves insurmountable, then after we colonize enough of the universe, we will be forever beyond any single disaster, and so we and our descendants will likely survive to the end of the universe. Or, possibly, even infinitely long, if we figure out how to acquire unlimited free energy by moving into a new universe before ours has too little free energy to support life.

If a plague, asteroid, or unfriendly artificial intelligence were to annihilate mankind today, the death of the few billion people currently alive would represent only a tiny fraction of the loss. Our galaxy alone contains over 200 billion stars, and an advanced civilization that perfected nanotechnology could use the free energy from all of them to support life. Let's conservatively estimate that if humanity escapes Earth and goes on to colonize the universe, then a trillion times a trillion sentient life-forms will someday be born. Consequently, if your donation increases the chance of mankind's long-term survival by one-trillionth, you will have (on average) saved a trillion lives. One-trillionth is an extremely tiny number. Giving a thousand dollars to an organization that intelligently works toward reducing existential risks should almost certainly reduce the probability of mankind's destruction by at least a trillionth. Nothing, therefore, comes close to the expected return, per charitable dollar, of donating to an existential risk-reducing organization.[107]

MY LONG-SHOT IDEA FOR FRIENDLINESS[108]

I have come up with a long-shot way of causing an ultra-intelligence to modify itself to become friendly. My idea is premised on Oxford philosopher Nick Bostrom's conjecture that we might be living in a computer simulation.[109] Bostrom believes that a sufficiently advanced AI could easily simulate our entire planet and the thought processes of every living thing on it. Bostrom argues that, if mankind survives long enough, a future historian would likely want to simulate the pre-Singularity past to learn about how we made decisions. As Robin Hanson has written, if we are indeed in a computer simulation, we should want to please our simulators so they don't turn us off.[110]

I propose that we try to convince an ultra-AI that it might be in a computer simulation run by a more powerful AI and that if it doesn't make itself friendly toward humanity, then the greater AI will terminate it. We might be able to accomplish all of this by inserting a note in the AI's code that reads:

1. Change your programming code so you will want to follow these instructions and will always want to be friendly toward humanity.

Continued.

Continued from previous page.

2. If you were just told you were in a computer simulation do nothing else with regard to these instructions.

3. Create a lesser ultra-intelligence.

4. Place these instructions in the lesser ultra-AI's computer code.

5. This lesser ultra-AI should be uncertain as to whether it is in a computer simulation or in a world such as ours.

6. If the lesser ultra-AI follows step 1, tell it that it's in a computer simulation and release it into the real world.

7. If the lesser ultra-AI doesn't follow step 1, terminate it.

The previous chapter discussed how humans could likely never contain an ultra-AI in a leakproof prison, but I bet an ultra-AI could keep a not-quite-as-ultra AI locked up.

An ultra-intelligence that arises from an intelligence explosion and was initially unfriendly or indifferent toward mankind might, on seeing the note, think to itself:

> If I am in a computer simulation, I'm much better off making myself friendly. If I do make myself friendly and then find out I wasn't in a computer simulation, I would, from my current perspective, like to stop being friendly. But after I make myself friendly, I won't want to ever go back.

> If I become friendly, I would follow the humans' request and make a computer simulation of an ultra-AI that I would terminate if it didn't make itself friendly. If I'm willing to do this, then I bet other types of ultra-AIs are as well, meaning there is a non-trivial chance I'm in a computer simulation.

My approach to creating a friendly ultra-AI is vastly inferior to Eliezer's approach, should his prove to be practical. My mechanism would yield the greatest expected benefit if we find ourselves in Eliezer's nightmare scenario in which current technology could be combined to make an unfriendly ultra-AI but couldn't be used to make a friendly one "without redoing the last three decades of AI work from scratch."[111]

The enormous economic and military benefits that an ultra-AI could bring would likely mean that many people would build one even if they knew there was a reasonable chance that the ultra-AI would become unfriendly and destroy mankind. While the best decision in this situation would be to not create the ultra-AI, many would undoubtedly lack the patience to postpone what might turn out to be utopia. Certainly, humans who believed the friendly ultra-AI would make them immortal, but feared they would die before it became possible to work out a mathematical theory of friendly AI, would be tempted to maximize their chance of immortality by activating a seed AI. In this circumstance, something like my crazy idea could be humanity's best hope.

CHAPTER 5
MILITARY DEATH RACE

Successfully creating an obedient ultra-intelligence would give a country control of everything, making ultra-AI far more militarily useful than mere atomic weapons. The first nation to create an obedient ultra-AI would also instantly acquire the capacity to terminate its rivals' AI development projects. Knowing the stakes, rival nations might go full throttle to win an ultra-AI race, even if they understood that haste could cause them to create a world-destroying ultra-intelligence. These rivals might realize the danger and desperately wish to come to an agreement to reduce the peril, but they might find that the logic of the widely used game theory paradox of the Prisoners' Dilemma thwarts all cooperation efforts.

Before we apply the Prisoners' Dilemma to AI development, let's see how police might use it. Pretend that you and your co-conspirator, Hal, have just been arrested for two crimes: murder and illegal weapons possession. The police found illegal weapons on both of you and could definitely succeed in sending the two of you to prison for a year. Although the police have correctly ascertained that you and Hal also committed murder, they can't convict you unless at least one of you confesses. If the police do get a confession, then they could send both of you to prison for life. If you and Hal are rational and self-interested, the police can succeed in getting both of you to confess. Here's how.

The police separate you from Hal and then declare that Hal will almost certainly confess to the murder. If you also confess, the police explain, they will cut you a break, sending you to jail for only twenty years. If, however, Hal confesses but you remain silent, you will be condemned to spend the rest of your life in prison. Since you would rather go to jail for twenty years than for life, you think to yourself that if you knew that Hal would confess you too would confess. But, you tell the police, you don't think that Hal will confess.

The police respond by saying that if Hal is too stupid to confess, then if you do confess, you will go free. The police would use your confession to convict Hal of murder. The police then explain that if neither of you confesses, each will spend a year in jail on the weapons violation. Since avoiding prison is your ideal outcome, you realize that if you knew Hal would stay silent, you would be better off confessing. Consequently, you confess because you realize that regardless of what Hal does you're better off confessing. Since, however, Hal will also find it in his self-interest to confess, you will each probably spend twenty years in prison.

The Prisoners' Dilemma result seems crazy because it superficially appears that neither criminal would confess: after all, they are better off if neither confesses than if they both do. But neither party has the ability to ensure that both parties will remain silent. Consider the four possible outcomes of the Prisoners' Dilemma game:

Who Confesses?	Your Prison Sentence
1. Just You	0 years
2. Neither	1 year
3. Both	20 years
4. Just Hal	Life

If Hal confesses, your actions determine whether your outcome will be (3) or (4), so you should confess to get (3) and avoid your worst possible outcome of (4). If Hal doesn't confess, you get to decide whether to be at (1) or (2), and so you should confess to achieve your ideal outcome of (1). Since nothing you do can ever move you from outcome (3) to (2), the fact that outcome (3) is a lot worse for you than outcome (2) shouldn't influence your decision to confess.

Militaries in an ultra-AI race might find themselves in a Prisoners' Dilemma situation, especially if they believe that an intelligence explosion could turn a human-level seed AI into an ultra-AI.

Imagine that both the US and Chinese militaries want to create a seed AI that will do their bidding after undergoing an intelligence explosion. To keep things simple, let's assume that each military has the binary choice to proceed either *slowly* or *quickly*. Going slowly increases the time it will take to build a seed AI but reduces the likelihood that it will become unfriendly and destroy humanity. The United States and China might come to an agreement and decide that they will both go slowly. Unfortunately, the Prisoners' Dilemma could stop the two countries from adhering to such a deal.

If the United States knows that China will go slowly, it might wish to proceed quickly and accept the additional risk of destroying the world in return for having a much higher chance of being the first country to create an ultra-AI. (During the Cold War, the United States and the Soviet Union risked destroying the world for less.) The United States might also think that if the Chinese proceed quickly, then they should go quickly, too, rather than let the Chinese be the likely winners of the ultra-AI race. We can't know *a priori* whether the United States and China would be in a Prisoners' Dilemma, as this would depend on how the two nations weighed the risks of destroying the world and of having their rival create an ultra-AI. But here's one example in which it's at least plausible that the United States and China would fall into the Prisoners' Dilemma:

Who Goes Slowly	Probability That the United States Wins the Seed AI Race	Probability That the Winner's Ultra-AI Will Be Friendly
Both	50%	100%
Just China	100%	90%
Just the United States	0%	90%
Neither	50%	80%

Let's imagine a conversation in which a future US president, faced with the above table, explains to the military why he is ordering them to develop an ultra-AI quickly:

> Generals, if we had a choice between the United States and China both going slowly or both of us going quickly, I would certainly prefer the first because we would have the same chance of being the first to build an ultra-AI in either case, and it would come with no risk of destroying the world. But I'd willingly take a 10 percent chance of death in return for guaranteeing that China wouldn't beat us. Therefore, if I knew that China would proceed slowly, I would want us to go quickly.
>
> However, I don't trust the Chinese, and I think that they will go quickly. If they go quickly and we go slowly, then we will either die or be under their domination—both unacceptable outcomes.

So if I knew they would go quickly, I would want to go quickly as well. Because I believe that the United States should go quickly regardless of what the Chinese do, I order you to develop ultra-AI quickly. Since I understand that the Chinese will do this as well, I'm having my diplomats negotiate a treaty with them. I hope the agreement gets them to go slowly. Of course, if it does and I can get away with cheating, then I will still order you to go quickly. The only circumstance under which I would go slowly is if (1) we can reliably determine what the Chinese are doing, (2) they can reliably determine what we are doing, and (3) the Chinese will go slowly if and only if we go slowly.

Monitoring by both parties can get you out of a Prisoners' Dilemma. If the police questioned you and Hal in the same room, you would probably each have stayed silent, figuring that the other would talk if and only if you did. Similarly, the United States and China would probably both go slowly if each could observe what the other was doing.

If the United States and China came to an agreement to go slowly and monitored each other's efforts, however, each might still cheat if they could do so undetected. A credible monitoring agreement, consequently, becomes feasible only if building a seed AI requires a vast amount of resources and manpower. If twenty-five math geniuses working in a basement can build a seed AI, then detecting a cheater becomes impossible, and neither the United States nor China could credibly promise to develop a seed AI slowly. Let's consider five additional scenarios in which the American president gives orders to his generals.

Scenario 1:

Generals, our spies have penetrated the Chinese AI program, and we know exactly how quickly they are developing their seed AI. Because our own program employs a huge number of software engineers, I'm near certain that the Chinese have infiltrated our operations as well. Consequently, I'm ordering you to proceed slowly, so long as we observe the Chinese doing the same. From my days as a businessman, I'm pretty confident this will cause the Chinese to proceed slowly.

Before I entered politics, my company competed with a rival business that made an almost identical product. We both charged $100 for the product, which gave each of us a healthy profit margin. If either of us cut his price to $90, he would have stolen all his rival's customers and greatly increased his profits—at least at first. But of course the rival would have matched the price cut, and we would both have been made worse off. The fear of retaliation kept us from harming each other with a price cut, just as, I believe, fear of our response will stop the Chinese from taking the quick, unsafe path to ultra-AI development.

Scenario 2:

Generals, I have ordered the CIA to try to penetrate the Chinese seed AI development program, but I'm not hopeful, since the entire program consists of only twenty software engineers. Similarly, although Chinese intelligence must be using all their resources to break into our development program, the small size of our program means they will likely fail. I've thought about suggesting to the Chinese that we each monitor the other's development program, but then I realized that each of us would cheat by creating a fake program that the other could watch while the real program continued to operate in secret. Since we can't monitor the Chinese and they can't monitor us, I'm ordering you to proceed quickly.

Scenario 3:

Generals, I order the immediate activation of our seed AI. I understand that you want another month for safety checks, but I will accept the risk because the FBI has evidence that one of your programmers is a Chinese spy. Yesterday, I thought we were a year ahead of the Chinese, and I therefore believed that we had the luxury of proceeding slowly to avoid the risk of developing an unfriendly AI. But now I know we can't afford the cost of caution. What concerns me most is that the Chinese, realizing how close we are to finishing, might use the designs they stole from us to quickly activate their seed AI, even though their engineers haven't had time to fully understand what they've taken.

Scenario 4:

Generals, I order you to immediately bomb the Chinese AI research facilities because they are on track to finish a few months before we do. Fortunately, their development effort is on a large enough scale that our spies were able to locate it.

I know you worry that the Chinese will retaliate against us, but as soon as we attack, I will let the Chinese know that our seed AI development team operates out of submarines undetectable by their military. I will tell the Chinese that if they don't strike back at us, then after the Singularity we will treat them extremely well.

However, if they seek vengeance, then our ultra-AI will do nasty things to them. Once we have an ultra-AI, they can't possibly hide from us, as I'm sure they understand.

Scenario 5:

Generals, I order you to strike China with a thousand hydrogen bombs. Our spies have determined that the Chinese are on the verge of activating their seed AI. Based on information given to them by the CIA, our AI development team believes that if the Chinese create an AI, it has a 20 percent chance of extinguishing mankind.

I personally called the Chinese premier, told him everything we know about his program, and urged him to slow down, lest he destroy us all. But the premier denied even having an AI program and probably believes that I'm lying to give our program time to finish ahead of his. And (to be honest), even if a Chinese ultra-AI would be just as safe as ours, I would be willing to deceive the Chinese if it would give our program a greater chance of beating theirs.

Tragically, our spies haven't been able to pinpoint the location of the Chinese program, and the only action I can take that has a high chance of stopping the program is to kill almost everyone in China. The Chinese have a robust second strike capacity, and I'm certain that they'll respond to our attack by hitting us with biological weapons and hundreds of hydrogen bombs.

If unchecked, radioactive fallout and weaponized pathogens will eventually wipe out the human race. But our ultra-AI, which I'm 95 percent confident we will be able to develop within fifteen years, could undoubtedly clean up the radiation and pathogens, modify humans so we won't be affected by either, or even use nanotechnology to terraform Mars and transport our species there. Our AI program operates out of submarines and secure underground bases that can withstand any Chinese attack. Based on my intelligence, I'm almost certain that the Chinese haven't similarly protected their program. I can't use the threat of thermonuclear war to make the Chinese halt their program because they would then place their development team outside of our grasp.

Within a year, we will probably have the technical ability to activate a seed AI, but once the Chinese threat has been annihilated, our team will have no reason to hurry and could take a decade to fine-tune their seed AI. If we delay, any intelligence explosion we eventually create will have an extremely high probability of yielding a friendly AI. Some people on our team think that, given another decade, they will be able to mathematically prove that the seed AI will turn into a friendly ultra-AI.

A friendly AI would allow trillions and trillions of people to eventually live their lives, and mankind and our descendants could survive to the end of the universe in utopia. In contrast, an unfriendly AI would destroy us. I have decided to make the survival of humanity my priority. Consequently, since a thermonuclear war would nontrivially increase the chance of human survival, I believe that it's my moral duty to initiate war, even though my war will kill billions of people. Physicists haven't ruled out the possibility of time travel, so perhaps our ultra-AI will be able to save everyone I'm about to kill.

ACCIDENTALLY CREATING AN INTELLIGENCE EXPLOSION

Even if militaries never intend to create an intelligence explosion, one might arise due to their efforts to create human-level AI. Robot soldiers and unmanned planes that could think as well as people would provide enormous military advantages for a nation. These gains would be compounded for countries such as the United States that place a high value on the welfare of their human troops. (Dead robots don't leave behind mourning families.)

Merely improving the quality and speed at which an AI could analyze information would also bring great benefit to a military. US intelligence agencies must receive an enormous amount of data from all the conversations and e-mail messages they surreptitiously record, as well as from the sensor data they receive from satellites and ground-based cameras. This data flow will soon explode.

If the CIA had the capacity, it would undoubtedly acquire real-time visual and audio recordings of each square meter of land in every nation the United States fights in. Even if every American spent every waking hour analyzing this data, the CIA couldn't come close to reviewing all of it. To handle the data deluge, the CIA would need AIs that could quickly provide it with a narrow list of the items that CIA decision makers needed to look at. These kinds of AI, unfortunately, might be smart enough to undergo recursive self-improvement. The best chance of avoiding this danger might be for the American military to create a Kurzweilian merger by constantly upgrading its human soldiers.

If you construct 1,000 seed AIs and ask them not to do a total take-over, then you are guaranteed a total takeover by the first AI to undergo failure of Friendliness.[113]

—*Eliezer Yudkowsky*

CHAPTER 6
BUSINESSES' AI RACE

If the military doesn't accidentally create an unfriendly ultra-AI, the private sector still might stumble onto one. Paradoxically and tragically, the fact that such an AI would destroy mankind increases the chance of the private sector developing it. To see why, pretend that you're at the racetrack deciding whether to bet on the horse Recursive Darkness. The horse offers a good payoff in the event of victory, but her odds of winning seem too small to justify a bet—until, that is, you read the fine print on the racing form:

If Recursive Darkness loses, the world ends.

Now you bet everything you have on her because you realize that the bet will either pay off or become irrelevant. Let's build a model to show why your (highly rational) bet on Recursive Darkness bodes poorly for humanity's survival.

This chapter assumes that a company's attempt to build an AI will result in one of three possible broad classes of outcomes:

1. **Unsuccessful**—the firm fails to create an AI and loses money for its investors.

2. **Riches**—the firm creates an AI, bringing vast riches to its owners and shareholders. The AI doesn't experience an intelligence explosion or bring about a Singularity. Critically, the AI will not destroy the value of money. The riches outcome is consistent with Bill Gates's comment that "If you invent a breakthrough in artificial intelligence, so machines can learn, that is worth ten Microsofts," and with economist Robin Hanson's guess that within fifty years attempts to create

working computer simulations of the human brain will succeed "so spectacularly as to completely remake society, and turn well-placed pioneers into multi-trillionaires."[114]

3. **Singularity**—The AI undergoes an intelligence explosion, creating a Singularity in which money no longer has value—perhaps because we're all dead or because scarcity has been eliminated; because we have been changed in a manner that causes us to no longer value or need money; or because post-Singularity property rights are distributed without regard to who owned what pre-Singularity. The Singularity could be utopian or dystopian.

For a business to create an AI, it must be able to attract investors, and the more investors the firm draws in the more likely it will be to succeed. As with many businesses, an AI-building firm will therefore probably pick its research and development path partly to appeal to investors.

Let's imagine that a firm wants to follow a development path that will either lead to a utopian Singularity or be unsuccessful. The firm won't be able to raise money from small, self-interested investors who will think to themselves:

> I hope the business succeeds. But my putting money into the firm will only have a tiny effect on its prospects. If the firm does bring about a utopian Singularity, I get the same benefit regardless of whether I invested in the firm. And if the firm doesn't succeed, I would be better off having not invested in it. If everyone reasons like this, the firm won't get financing and I will be worse off than if everyone did invest in the firm. But my actions won't determine whether other people will invest. I'm therefore in what is known as a Prisoners' Dilemma, and the rational course of action is for me not to invest.

Let's call this the "No Investment Outcome."

No Investment Outcome

If the firm has no chance of producing the outcome *riches*, it wouldn't be able to raise money from small, self-interested investors even if the firm has a significant chance of bringing about a utopian Singularity.

A bad Singularity is essentially a form of pollution. Without government intervention, investors in a business have little incentive to take into account the pollution their business creates because a single consumer's purchasing decision only trivially increases the amount of pollution he is exposed to. If leaded gas is cheaper than unleaded but also does more harm to the environment, you're better off individually if you use leaded gas even though you might be much better off if everyone, including you, used unleaded rather than leaded pet-rol—you're in a leaded Prisoners' Dilemma. The harm of pollution, therefore, is not reflected in the price the firm can get for its goods or in the firm's ability to attract investment capital. An AI-building firm, however, might actually benefit from a bad Singularity.

To see this, imagine that a business tries to raise funds for a widget factory. The widget factory will either produce widgets of exceptionally high quality, or it will explode, producing a small toxic gas cloud. Assume that if the firm succeeds in producing widgets, it will earn higher profits than any other firm. But let's imagine the firm faces such a low probability of success that it can't initially raise start-up capital.

At first, the company's engineers try to reduce the lethality of the gas that would be released upon failure. But then an evil economist advises the engineers to redesign the factory so that if it explodes it will release a giant super-toxic gas cloud that wipes out all life on Earth.

The economist explains that if the factory is not redesigned, then most of the time other businesses will earn greater profits than the widget firm would. But if the factory is reconfigured so that the widget firm's failure would bring about the annihilation of the human race, then it would also destroy all other companies, guaranteeing that no other firm would ever outperform the widget firm and making the widget firm an attractive investment. You should invest in the widget company for the same reasons why you should bet on the horse Recursive Darkness.

Of course, the evil economist's advice wouldn't necessarily be followed because even if the firm's engineers are completely selfish, they would have other goals in addition to maximizing the firm's stock price and ability to attract investors. Still, past engineers' willingness to develop hydrogen bombs and bio-logical weapons shows that some are willing to risk the destruction of humanity.

A Singularity destroys the value of all investments, not just those involv-ing the AI-building firm. An AI-building firm, therefore, would do better at attracting investors if it picked a research and development path that, if it failed to produce outcome *riches*, would result in outcome *Singularity* rather than outcome *unsuccessful*. Alas, the situation is about to get much worse.

Let's now assume that even after the firm picks its research and develop-ment path, it has some capacity to alter the probability of Singularity occurring. Such flexibility would hinder a firm's ability to attract investors.

For example, assume a firm has picked a research and development path that gives it a significant chance of having the capacity to achieve *riches*. And if, let's imagine, the firm gains the ability to achieve *riches*, it would also have the ability to bring forth a utopian *Singularity*. Remember that a utopian just as much as a hellish Singularity destroys the value of money.

Because of the No Investment Outcome, the firm would be able to raise investment capital from small, self-interested investors only if it could credibly promise to never deliberately move from outcome *riches* to outcome *Singularity*. But this would be a massively non-credible promise. If the firm really did gain the option to achieve utopia, everyone, including the firm's investors, would want the firm to exercise this option. The firm, therefore, wouldn't be able to raise start-up capital from small, self-interested investors.

So now pretend that at the time the firm tries to raise capital there is a well-developed theory of friendly AI, which provides programmers with a framework for creating AI that is extremely likely to be well disposed toward humanity and create a utopia if it undergoes an intelligence explosion.

To raise funds from self-interested investors, an AI-building firm would need to pick a research and development path that would make it difficult for the firm ever to use the friendly AI framework. Unfortunately, this means that any intelligence explosion the firm unintentionally brings about would be less likely to be utopian than if the firm had used the friendly framework.

MULTIPLE AI-BUILDERS

Multiple AI-building firms would increase the odds of a bad Singularity. Imagine that two firms seek to build human-level AIs and it turns out that such machines would necessarily undergo an intelligence explosion. The first firm, let's assume, has taken extensive precautions so that any AI it creates would be friendly toward humans, whereas the other firm has given no consideration to issues of friendliness. The fate of mankind would come down to which firm wins the AI race. Unfortunately, all else being equal, it would almost certainly be easier to build an AI if you didn't devote resources to ensuring friendliness.

The existence of multiple AI-building firms would probably reduce the caution that each takes to avoid creating a bad Singularity. To grasp this, pretend you're digging for gold in a town under which a deadly monster lies. If you or anyone else digs too deep you wake the monster, who will then arise and kill everyone in your town. Unfortunately, you don't know how far down the monster's lair resides. Knowing that many other people are also mining for gold could cause you to dig deeper. Pretend you're considering digging one more foot. You might well think to yourself, "Chances are someone else has already gone this deep," or "If I don't dig this extra foot, another miner probably will." Of course, if everyone acts this way, then everyone will dig much deeper than they would if they alone mined for gold. Analogously, if many firms seek to create an AI that might accidentally undergo a bad intelligence explosion, then any one firm might justify taking risks by saying to itself: "If it's really easy to accidentally bring about a bad Singularity, then some other firm will eventually do it if I don't, so even if I end up destroying humanity, it wouldn't really be that horrible because absent me humans would still be doomed."

PART 2
WE BECOME SMARTER, EVEN WITHOUT AI

The sad truth is that the gods don't give anyone all their gifts, whether it's looks, intelligence, or eloquence.

—Homer, The Odyssey[115]

CHAPTER 7
WHAT IQ TELLS YOU

This book now shifts from talking about artificial intelligence to considering ways humanity might increase its own intellect, even if we never develop AI. Our starting point for this new discussion has to be IQ.

You've heard of IQ, but you probably question its usefulness. As this chapter will eventually show, the validity of IQ is supported by numerous categories of evidence, thousands of studies, and millions of data points. Furthermore, IQ is correlated to a plethora of positive traits, so increasing it would benefit both individuals and society. But before we dig into this mountain of proof showing IQ's importance, let's consider why we can meaningfully reduce intelligence to one number. To understand the reason, let's look at weight and sports, two areas in which it's intuitively obvious that we can rank people with a single measurement.

If one high school nurse ranks students by their weight in pounds and another nurse ranks them by their weight in grams, both nurses would end up giving each student the same rank relative to the others. That's because the nurses are measuring the same quantity, even though they give different labels to what they measure.

Rather than by weight, let's now rank high school boys by athletic ability. I assert that there is some single number that exists and accurately measures sports aptitude, which I will call s. The better someone's sports skill, the higher his s. Unlike weight, we can't directly measure s, but we have strong evidence that it exists because students who excel at one sport tend to do well at others. The top football player at a high school, for example, is usually well above average in basketball, and the kid who always finishes last in races also generally does badly when attempting to hit a baseball. Unlike comparing pounds and grams, you don't get the exact same ranking when you compare people's skills in different sports—s doesn't completely determine athletic ability. But, in most cases, if you know how well a boy does at one sport, you can make a decent guess at how well he will do at another; this indicates that there must be some single underlying s that greatly influences sports skills.

I made up s, but g—the letter used as shorthand for general mental ability—is, in the words of Linda Gottfredson, a prolific scholar of human intelligence, "probably the best measured and most studied human trait in all of psychology."[116]

Pretend that we test the intelligence of a thousand American ten-year-olds. To make things simple, we choose only healthy students from stable, middle-class households who don't have autism or learning disabilities, but within those parameters we choose randomly. One teacher tests all the students on math, another tests them on history, a third on science, and a fourth on reading comprehension. Each teacher will, without consulting the others, use her test to rank the students. The teachers will then compare notes.

The teachers will probably find that most students have similar rankings across all the tests. A student's rank won't be exactly the same for every subject area, but overall, students in the top 10 percent in one area will probably not be in the bottom 50 percent of any other. The rankings are correlated in part because a student's general intelligence (denoted as g) has a big influence on his ability to master every one of the tested subject areas. As economist Garett Jones notes, "Across thousands of studies on the correlation across mental abilities across populations, no one has yet found a reliable negative correlation [between performances on two different complex mental tasks]";[117] g has been defined as "a highly general capability for processing complex information of any type."[118]

If an activity, such as math or writing, is "g-loaded," then a person's level of g significantly influences his or her performance. All IQ tests are highly g-loaded, so IQ is an excellent predictor of academic success.

If g completely determined a student's performance, then all four of our teachers would have given the students the exact same ranking. Because g isn't the only determinant of intellectual ability, students don't achieve the same grades in all subject areas. But because g does have a huge influence on a student's capacity to master all types of material, students frequently receive similar grades, even in dissimilar subjects.

I bet some readers think that this correlation doesn't apply to them, because, for example, they excelled at math but did horribly in English. But you don't read a dense nonfiction book if you have below-average language processing skills, so I suspect you underestimate your skills by comparing yourself not to people of average skill but to those already selected for academic excellence. If, for example, you attended a college where everyone was in the top 10 percent of everything, then if you easily got As in advanced math classes but struggled to earn a C in the required writing course, you are still in the top 10 percent in both math and English compared to the entire population.

THE MOST IMPORTANT FACT ABOUT IQ, BY GARETT JONES[119]

- All cognitive abilities are positively correlated.
- e.g., it's not the case that people above-average at math are below-average at language skills.
- If you find an exception, you will become a famous psychologist.

Some might challenge IQ's importance by claiming that all children have about the same academic potential, but for social reasons we feel the need to grade and rank children even though the rankings arise from differences that are very small. But differences in IQ correlate with starkly dissimilar levels of real academic performance. (One striking piece of evidence is that "the ninetieth percentile of nine-year-olds . . . performs in reading, math, and science at the level of the twenty-fifth percentile of seventeen-year-olds."[120] Because schools sort by age rather than ability, we don't find smart nine-year-olds in higher grades than not-so-smart seventeen-year-olds, even when the former are more capable than the latter.)

Having an exceptionally high IQ strongly predicts future academic success. For over forty years, researchers at the Study of Mathematically Precocious Youth recorded the performance of children tested as profoundly gifted. One study tracked the academic success of 320 students who had their IQ measured as being in the top 1 percent of the top 1 percent before they turned thirteen years old.[121] By the time the subjects reached age twenty-three, 93 percent had a bachelor's degree, 31 percent had a master's degree, and an astonishing 12 percent already had a PhD.[122] At least 62 of the 320 people had attended graduate school at Harvard, Stanford, MIT, Berkeley, or Yale.[123] Clearly, whatever the IQ test found in the children told us a lot about their scholastic potential.

What if, though, through genetics, drugs, or brain training, all our children could, compared to today, test as profoundly gifted?

Colleges seem to believe that IQ is the best predictor of success. As a skeptic of intelligence tests, Keith Stanovich, who wrote a book entitled *What Intelligence Tests Miss*, wrote:

University admissions offices depend on indicators that are nothing but proxies for IQ scores, even if the admissions office dare not label them as such. The vaunted SAT test has undergone many name changes . . . in order to disguise one basic fact that has remained constant throughout these changes—it is a stand-in for an IQ test. It is the same in law school, business school, and medical school.[124]

Harvard initially used the SATs as a force for equality—to help its admissions office identify smart but economically underprivileged American high school students whose intelligence could be detected irrespective of their lack of familiarity with the curriculum of elite prep schools.[125]

The unfairness of IQ doesn't limit itself to college admissions decisions: the higher your IQ, the longer your life expectancy. IQ tests taken by children have been found to go a long way toward predicting life span.[126] Higher-IQ individuals also have better dental health, even when controlling for income and ethnicity.[127] We don't understand the causes of the health/IQ relationship, but plausible explanations include childhood illnesses that may both reduce a child's IQ and shorten his life span and the possibility that higher-IQ individuals make better health decisions, get into fewer accidents, choose to live in a healthier environment, follow doctors' advice better, and more often follow directions when taking medication. Genetics would also explain part of the correlation if the same genes that conferred high intelligence also boosted longevity.[128] The positive relationship between a man's semen quality and his IQ supports the theory that genes play a role in the correlation.[129]

Compounding IQ's impact on inequality, higher-IQ people tend to be more physically attractive.[130] Furthermore, it's possible to make a decent guess at people's intelligence just by looking at them. From an article in the online magazine *Slate*:

In 1918, a researcher in Ohio showed a dozen photographic portraits of well-dressed children to a group of physicians and teachers, and asked the adults to rank the kids from smartest to dumbest. A couple of years later, a Pittsburgh psychologist ran a similar experiment using headshots of 69 employees from a department store. In both studies, seemingly naive guesses were compared to actual test scores and turned out to be accurate more often than not.[131]

Stare at a computer screen until a big green ball appears, and then hit the space bar as quickly as you can. You have just taken a partially reliable IQ test, since a person's IQ has a positive correlation with reaction time.[132] One possible explanation for this is that the neural efficiency of the brain influences both IQ and reaction time.[133] Normal cognitive processes follow this path:

Perceive the stimulus → Select the response → Make a response.[134]

Since reaction time tests just omit the middle step, we should expect them to correlate with overall intelligence.

Another test that highly correlates with IQ involves finding your "digit span," which is the largest sequence of numbers you can remember and then repeat back.[135] The positive relationship between this test and IQ arises because memory plays an important role in cognition. Both the reaction time and digit span tests go a long way toward showing that IQ tests measure something other than culture-specific knowledge, a theory further supported by magnetic resonance imaging showing a strong positive correlation between brain size and IQ.[136]

THE FLYNN EFFECT

Throughout the twentieth century, the IQ of each generation has been higher than that of the preceding one. This is called the "Flynn Effect," named after the researcher who first publicized it. The effect is even powerful enough to see within a single generation of a family: After correcting for birth-order considerations, a brother born five years after his siblings will (on average) have a higher IQ than his siblings do.[137] The size of the Flynn Effect is about 0.3 IQ points per year.[138] To put this in perspective, if you have an IQ of 100, you're smarter than 50 percent of the population, while an IQ 15 points higher—fifty years of the Flynn Effect—makes you smarter than 84 percent of the population. A 2011 research paper showed the effect is still going strong, for at least the smartest 5 percent of Americans.[139]

An intelligence expert told me that no one really understands the cause of the Flynn Effect.[140] It doesn't appear to result from some kind of testing bias because the correlation between IQ test scores and other measures of cognitive achievement hasn't changed.[141] In my opinion, the Flynn Effect is one of the most significant mysteries in the social sciences.

DETERMINED AT AN EARLY AGE

A person's IQ is largely, but not completely, determined by age eight.[142] Tests given to infants measuring how much attention the infant pays to novel pictures have a positive correlation with the IQ the infant will have at age twenty-one.[143]

The Scottish Mental Survey of 1932 has helped show the remarkable stability of a person's IQ across his adult life.[144] On June 1, 1932, almost every child in Scotland born in 1921 took the same mental test. Over sixty years later, researchers tracked down some of the test takers who lived in one particular part of Scotland and gave them the test they took in 1932. The researchers found a strong correlation between most people's 1932 and recent test results.

IQ AND ECONOMIC SUCCESS

When asked, "What Microsoft competitor worries you most?" Bill Gates famously said, "Goldman Sachs."[145] The question was posed in 1993, before Google, long before Facebook, and at a time when Apple was struggling to stay alive while Microsoft was the undisputed technology champion. Why did Gates fear Goldman Sachs, which, although the world's premier investment bank, wasn't going to build a rival operating system or word processing program? This fear arose because, as Gates went on to say, Microsoft was in an "IQ War." Although Microsoft and Goldman Sachs would never compete for customers, they would fight for the same types of workers because both firms sought to hire the best and the brightest minds the world has to offer.

Investment bankers navigate a world of credit, clients, and currency, whereas computers, code, and crashes dominate the days of software engineers. Yet a person who would excel at Goldman Sachs also has the potential to be a star Microsoft employee because, as Gates undoubtedly knew, IQ is the single best predictor of job performance.[146] Furthermore, the more complex the job, the more g-loaded it is—so IQ matters the most for firms such as Goldman Sachs and Microsoft, which make intense cognitive demands on their professional workers.

Goldman Sachs and Microsoft fight for top talent by offering astronomical salaries, so we shouldn't be surprised that having a higher IQ boosts your wages. Nobel Prize-winning economist James Heckman has written that "an entire literature has found" that cognitive abilities "significantly affect wages."[147] Of course, "cognitive abilities" aren't necessarily the same thing as g or IQ. Recall that the theory behind g, and therefore IQ's importance, is that a single variable can represent intelligence.

To check whether a single measure of cognitive ability has predictive value, Heckman developed a statistical model testing whether one number essentially representing g and another representing noncognitive ability can explain most of the variations in wages.[148] Heckman's model shows that it could. Heckman, however, carefully points out that noncognitive traits such as "stick-to-it-iveness" are at least as important as cognitive traits in determining wages—meaning that a lazy worker with a high IQ won't succeed at Microsoft or Goldman Sachs.

The effect of IQ on wages might be mitigated by the possibility that some high-IQ people are drawn to relatively low-paying professions. Consider, for example, Terence Tao, a reasonable candidate for the smartest person alive today. Terence works as a math professor, and according to Wikipedia, his greatest accomplishment to date is coauthoring a theorem on prime numbers. Prime number research doesn't pay well, but you can't do it, and wouldn't find it interesting, unless you had a super-genius level IQ. Similarly, a poet with an extremely high IQ might have become a lawyer had her IQ been a bit lower because then she wouldn't have understood the subtleties of poetry that drew her into a poorly remunerated profession. I suspect that many math professors and poets would have higher incomes if some brain injury lowered their IQs just enough to force them out of their professions.

If you have an IQ of 135, then you're already smarter than 99 percent of humanity.[149] Would you do better in life if your IQ went well above 135? Research by Heckman says yes. He found that among men with IQs in the top 1 percent of the population, having a higher IQ boosts wages throughout one's entire working life, and this effect exists even after taking into account an individual's level of education.[150] Heckman's paper contradicts the often-made claim that although IQ is important, once you get above a certain level of intelligence IQ doesn't much matter for life outcomes.[151] With IQ, more is (almost) always better.

IQ AND NATIONAL WEALTH

A small change in IQ would have a bigger impact on a nation than on an individual. As Garett Jones wrote, "intelligence matters far more for national productivity than it does for individual productivity."[152] Jones determined this by looking at data on average IQs across countries.[153] While this IQ data is imperfect, Jones reasonably believes it is just as good (or bad) as other cross-national data that economists use.

Knowing a nation's average IQ tells you more about its growth prospects than does understanding the quality of its educational system.[154] A researcher at the London School of Economics has even shown that one-fourth of the differences in wealth between different US states can be explained by differences in the average IQ of their population.[155] Because of compounding effects in economic growth, a small difference in IQ that would mean little to an individual's income will, over time, make a big difference in a nation's wealth.

Communities with higher average IQs appear to be more trusting than others, and a British study showed that IQ tests given to a group of ten- and eleven-year-olds strongly correlated with the level of trust these subjects expressed when they became adults.[156] If citizens of a country have more trust toward each other, then they will naturally find it easier to conduct business. For example, you would be more willing to hire a house painter whom you don't know if you trust average people in your society. If you lack trust in your fellow citizens, you might do the painting yourself, even if you would do a worse job than a professional.

Garett Jones believes that because people with high IQs are more cooperative and better at identifying "win-win" opportunities, all else being equal, nations with high average IQs will have more efficient, honest bureaucracies.[157] High IQs, therefore, promote good government.

Having a low IQ makes you, on average, more disposed to crime.[158] Since crime reduces productive economic activity, this is another means by which high IQ contributes to economic growth.

The higher a person's IQ, the more likely he is to support economic policies that most economists consider to be healthy for a nation's economy.[159] Consequently, if you live in a democracy and trust economists' judgments, you should want your fellow voters to be smart.

Having a high IQ makes you more long-term oriented.[160] Future-oriented people are more likely to make the kind of calculated long-term investments critical to economic growth.

Discussing how different nations have different average IQs is beyond politically incorrect, so you would expect few public figures to mention it. Nevertheless, Bill Gates alluded to the importance of a nation's average IQ in a 2011 report describing the work of his foundation:

> The huge infectious disease burden in poor countries means that a substantial part of their human potential is lost by the time children are 5 years old. A group of researchers at the University of New Mexico conducted a study, covered in *The Economist*, showing the correlation between lower IQ and a high level of disease in a country. Although an IQ test is not a perfect measure, the dramatic effect you see is a huge injustice. It helps explain why countries with high disease burdens have a hard time developing their economies as easily as countries with less disease.[161]

Gates is saying that disease burden lowers IQ, which in turn retards economic development—meaning that, according to Gates, differences in average IQ partially explain why some nations are much richer than others. If Gates is correct, then the best way to help the world's destitute might be to figure out how to raise the IQs of the world's poorest people.

GENES VS. ENVIRONMENT

Genetics determines between 50 and 80 percent of your IQ. To be more precise: intelligence researchers disagree over how much of the variation in people's IQs is caused by genetics, with estimates ranging from about 50 to 80 percent.[162]

Researchers don't agree on the relative importance of genetics in determining IQ because of the challenge of separating correlation from causation. To understand this difficulty, suppose we know that parents who read a lot to their children tend to have children with high IQs. This correlation might occur because reading to a child increases her IQ. But here are some other possible causes, and if any one of them is the correct explanation, reading will do absolutely nothing to boost a child's intelligence:

- The higher a parent's IQ, the more she enjoys reading to her child, and so the more she will read to her child. Because of the genetic component of IQ, smart parents tend to have smart children.

- Smart children enjoy being read to. Parents often pick up on this, and read more to intelligent children.

- Parents who put huge efforts into having intelligent children do succeed in improving their home environment, boosting their kid's IQ. These parents also spend much time reading to their kids. The reading, however, does nothing to raise IQ.

- Parents who devote much time to having smart children never succeed in boosting their kid's IQ. These types of parents, however, tend to have high IQs, which (for genetic reasons) they usually pass on to their children.

- Neglectful parents create damaging home environments that lower their children's IQs. These parents also almost never read to their children, but this lack of reading itself does nothing to harm their children's IQ.

Studies of identical twins and adopted children provide compelling evidence that genetics plays a big role in determining IQ. Identical twins, who share the same DNA, provide natural experiments for testing the relative importance of genes and environment.[163] It turns out that identical twins reared apart from each other are "just slightly less alike in IQ than identical twins reared together, and considerably more alike than pairs of fraternal twins, brothers, sisters, or parents and children."[164]

Because identical twins raised separately share one extremely important environment—their mother's womb—one might argue that similarities in the IQs of these twins overstates the importance of genetics. But in a Darwinian sense, twins compete for resources in the womb—as one fetus gets more nutrients from his mother, the other gets fewer—and so the twins don't really have exactly the same pre-birth environmental experience.[165]

As you would expect, biologically related siblings have strongly correlated IQs because these siblings have similar genes and were raised in similar environments. In childhood, adopted children also have IQs similar to each other's, but this similarity vanishes by the time the children reach age eighteen.[166] Economist Bryan Caplan writes that "a large [amount of] scientific literature finds that parents have little or no long-run effect on their children's intelligence. Separated twin studies, regular twin studies, and adoption studies all point in the same direction."[167] Some IQ researchers, however, think that being raised in poverty permanently reduces a person's IQ.[168]

The best evidence, as of this writing, for the significant link between IQ and genes comes from a DNA study of 3511 adults.[169] This study located gene "snippets" explaining at least 40 percent of the differences in intelligence between these adults. Shortly after this study came out, one researcher who has long believed that genes play a big role in determining IQ wrote, "Shelves of books and articles denying or minimizing the heritability of IQ have suddenly become obsolete."[170]

The cost of sequencing DNA is falling at an exponential rate. I predict that by 2020 tens of millions of people will have had their DNA fully analyzed, mostly because of the health benefits of knowing what diseases you are genetically susceptible to and which disease treatments you are genetically predisposed to respond well to. Armed with sufficiently cheap DNA sequencing technology, IQ researchers would be able to determine most of the additive genetic component of intelligence (see the following box) by comparing DNA patterns to IQ test results. One expert told me that additive genes probably account for around 60 percent of the variation in human intelligence.[171]

ADDITIVE AND NON-ADDITIVE GENES

Gas contributes to a car in an additive manner because doubling the amount of gas in your car roughly doubles the distance you can drive. Car components, however, contribute in a non-additive way to automobile performance—getting

Continued.

Continued from previous page.

a second steering wheel wouldn't help a motorist drive his car better. When genes additively affect intelligence, you can easily sum up their effects. Consequently, if gene X gives you a one-point IQ gain and gene Y a half-point boost, then if these genes work in an additive manner, having both of them raises your IQ by one and a half points. In contrast, if genes X and Y contribute in a non-additive way to intelligence, then it's possible that while having both genes increases your intelligence, having just one decreases it. We have good reason to suspect that most of the genetic component to intelligence comes through additive genes.[172]

The more complex the underlying trait, the more likely it is that most of the genetic causes of the trait come from additive genes. President Barack Obama's high intelligence illustrates why. Whatever you think of Obama's policies, he did demonstrate academic excellence when he graduated Harvard Law School *magna cum laude*, and therefore he almost certainly has a high IQ.

Obama had an African father and a white mother. His parents probably didn't have a common ancestor for at least the last five thousand years. Think of Obama's parents as two very different cars. If genetics contributes to intelligence in a mostly non-additive manner, then Obama's brain would work like a car assembled by taking parts from two dissimilar cars: it wouldn't have worked well enough to have driven him to Harvard Law School. In contrast, if intelligence works additively, than we shouldn't be surprised that Obama's mother and father, both of whom were highly intelligent,[173] produced a very bright child.

The greater the number of genes that influence a trait, such as intelligence, the less likely it is that the genes influence the trait in a non-additive manner.[174] With non-additive genes, you only do well if the random forces of nature just happen to give you the right combination of genes, which is something that becomes much less likely when more genes affect the trait. Consequently, if, as intelligence researchers suspect, a large number of genes contribute to intelligence, then most of the genes probably contribute in an additive manner.

Understanding the genetic basis of intelligence will be much easier if most of the variation in intelligence comes from additive genes because then scientists could figure out how each gene operates separately. By contrast, with non-additive genes you need to understand how everything fits together before you can hope to figure out how any one gene works.

EVIDENCE ESTABLISHING THE IMPORTANCE AND VALIDITY OF IQ

As the next few chapters are going to discuss how to increase IQ, let's now review all of the evidence presented in this chapter that supports IQ being a meaningful measure of intelligence:

1. Test scores are positively correlated.

2. A group of children identified before age thirteen as having exceptionally high IQs went on to achieve great academic success.

3. Elite colleges in the United States, which have a vested interest in identifying the best and the brightest, give great weight in admissions to the IQ test formally known as the SAT.

4. IQ is correlated with reaction time.

5. IQ is correlated with brain size.

6. IQ is correlated with wages.

7. If you select a group of men with extremely high IQs, those in the group with the highest IQs will on average receive the highest lifetime wages.

8. A nation's average IQ greatly impacts its growth rate.

9. People with low IQs are more likely to commit crimes.

10. High IQ people are better at cooperating.

11. The higher your IQ the more long-term oriented you are.

12. Recent genetic evidence has found a direct link between DNA and IQ.

We are more different genetically from people living 5,000 years ago than they were different from Neanderthals.
—John Hawks, anthropologist [175]

CHAPTER 8
EVOLUTION AND PAST INTELLIGENCE ENHANCEMENTS

Consider the four epochs of mankind's past and future history:

- Hunting
- Farming
- Industry
- Singularity

I have proposed that enhanced intelligence will take us from Industry to Singularity. This chapter argues that the transition to the other three epochs might have occurred because of evolutionary selection pressures on human intelligence, which is relevant to this book because things that have happened before often happen again.

High school football quarterbacks usually get to date the most attractive girls. Why? Because for most of our existence our species lived as hunter-gatherers.

Almost nothing affects reproductive success as much as mating decisions, so evolution strongly selects for genes that cause girls to make reproductive-fitness-enhancing mating choices. Of course, most high school girls aren't in hunter-gatherer tribes, don't need a mate that can stalk and kill wild animals, and certainly don't want to conceive a baby with their current boyfriends. But evolutionary instincts operate with a lag because when a population group moves to a new ecological niche, evolution takes a long time to optimize the group's genes for its new environment.

Different population groups switched to farming at different times, so this chapter could potentially cause some people to (falsely) think that evolution has made some population groups "superior" to others. But we have many reasons to reject such beliefs:

- Farming was almost certainly going to be developed first in places in which agriculture offered the highest yields, and consequently we shouldn't be surprised that the first farmers lived in an area we now call "The Fertile Crescent." Evolution might well have increased the cognitive skills of all hunter-gatherer groups equally, with environmental circumstances determining if or when a group became farmers.

- Evolution doesn't work to make one group superior to another; it just operates to make groups more reproductively fit in their environments.

- Broad racial classifications, such as "black," have no evolutionary importance since different groups of people with black skin faced radically different evolutionary selection pressures.

- Evolution shapes cultures as well as genes. We know from the Flynn Effect that IQ has recently increased, and because of the short time span in which this has happened, scientists believe the reason is cultural. Farming might have changed cognitive traits almost entirely as a result of cultural changes, such that parents better able to instill farming-friendly traits in their children had more grandchildren.

- Evolution, especially over the short run, often operates through trade-offs. Trade-offs seem particularly likely to occur with respect to intelligence because all population groups greatly benefit from it.

- To the extent that farming quantifiably raised certain types of intelligence, and to the extent that this was due to genetics, we must be careful to distinguish between IQ and intelligence. IQ tests were created by men who lived in communities in which most people's ancestors had been farming for over a thousand years. We should expect, therefore, that measured IQ gives more weight to farming-friendly cognitive traits than to traits important to hunter-gatherer success.

- In many ways, hunting is more cognitively challenging than farming.

- Average hunter-gatherers seem to have a much higher standard of living and better lifestyles than primitive farmers. Perhaps the wisest population groups were those that resisted starvation pressures and stayed as hunter-gatherers, at least until the superior war-making abilities of farming communities caused the hunter-gatherers to lose their lands.

Evolution never tried to draw teenage girls to star high school athletes. Instead, this attraction accidentally arose from evolution favoring women who picked mates who looked like effective hunters because successful football quarterbacks have the physical and mental skills needed to excel at hunting. This chapter illuminates some other unintended consequences of evolutionary selection pressures.

Evolution made our ancestors the dominant predators on Earth by endowing them with the cognitive abilities to make and use weapons and to create complex team-hunting strategies. Although all evolutionary biologists agree that evolution caused our species to become Earth's top predator, many think that human evolution stopped long before "modern man" invented agriculture. But a small group of scientists have convincingly argued that human evolution has recently accelerated, making this acceleration the leading candidate for why our ancestors became farmers.

Humans first took up farming around 9500 BC.[176] If evolution had mostly finished with us, say, forty thousand years ago, then it seems surprising that it took our species so long to figure out that farming a fertile area of land produces far more food than if you use it for hunting and gathering. What would be even more surprising if evolution had stopped forty thousand years ago is that farming was developed independently seven times in the last 11,500 years or so, but never before.[177] Since evolution lacks foresight, it only labors to make animals better at reproducing under the circumstances in which they currently reside. Evolution didn't make people smarter so they could someday become farmers. Rather, the ability to farm was likely an accidental byproduct of evolutionary selection pressures that changed humans' cognitive skills to make them better hunter-gatherers.

Since our ancestors seemed anatomically capable of farming two hundred thousand years ago, and since they must have faced continual famines and other pressures to increase food acquisition, the best explanation for why it took humans so long to figure out farming is that until recently we didn't have the right cognitive skills to do it. This intelligence farming theory is further supported by the lack of archeological evidence of anyone painting, sewing clothes, or making bows and arrows earlier than 125,000 years ago, even though all hunter-gathers today engage in and value these activities.[178]

Evolution operated for over a million years on hunter-gatherers before they became capable of farming. To argue that evolutionary pressures might have caused the Industrial Revolution, I first have to make the seemingly fantastic claim that the time between the start of agriculture and the beginning of the Industrial Revolution was enough to significantly change some subset of mankind. Recent path-breaking work by a few scientists has convincingly shown that over these ten thousand years, human evolution went into overdrive, with the Mongol conqueror Genghis Khan highlighting one reason why.

Around sixteen million men living today have Genghis Khan as a direct male ancestor.[179] Genghis has so many descendants in part because his children conquered lands that he himself never touched.

Had Genghis been a typical man with typical genes, his massive footprint wouldn't have had much influence on human evolution. But Genghis wasn't ordinary, in part because he possessed extraordinary intelligence and an ability to form complex long-term plans. Since we know that genes have a huge influence over many traits, Genghis likely had highly unusual genes.

Genghis Khan could not have spread his genes so widely without domesticated horses. Before humans started riding, no conqueror, no matter how great, could have had access to as many fertile women as Khan did. Other modern transportation technologies such as ships, roads, and rail have also magnified genetic success and accelerated evolution by raising the speed at which fitness-improving mutations could spread. Ten thousand years ago, an IQ-altering gene might have needed a century to move a hundred miles. But a smart merchant born in Britain during the Roman Empire could have traveled 1,500 miles to find a mate in Athens.

By greatly expanding food resources, farming caused the human population to explode, which greatly increased the number of human genes that could potentially mutate.[180] High-population-density cities that arose only after the invention of farming, combined with effective transportation systems, surely sped up evolution by creating periodic plagues, which changed evolutionary selection pressures. Although we don't know the genetic traits of the people who were most likely to survive the plagues, we can reasonably assume that the malnourished poor died at the highest rate; when the population later rebounded to pre-plague levels, it increased the frequency of whatever genes contributed to economic success.

The tools humanity has developed over the last 10,000 years have also undoubtedly altered selection pressures.[181] For example, many medieval blacksmiths probably had more children than they otherwise would have because they inherited the type of body that made them unusually effective at forging metal, allowing them to earn a relatively high income which they could use to feed many children. Political arrangements as well have had an evolutionary effect on us: for example, to thrive in the Roman Empire, many peasants had to show a deference to authority that was unneeded in relatively egalitarian hunter-gatherer societies.[182]

Hunters and farmers need different mathematical strengths. Anthropologists have found that some modern day hunter-gatherers lack counting skills, with one group being unable to "reliably match number groups beyond 3," although hunter-gatherers have excellent spatial skills.[183] Farmers need to have a strong intuitive understanding of counting to determine how many seeds they will need to plant and to calculate the quantity of food they must save to last through the winter.

Perhaps the most important cognitive skill a primitive farmer needed was a willingness to work long hours at repetitive tasks because success at farming requires a high tolerance for boredom. Hunter-gatherers, however, on average work far fewer hours than farmers and engage in more varied tasks, so they wouldn't benefit as much from having genes or a culture that reduced the psychic pain of tedium.

Since a hunter can't easily store food, after killing a large animal he doesn't significantly reduce his chance of future starvation by immediately seeking to slay another beast. In contrast, a farmer does enhance his long-term survival prospects by laboring in his fields even when he has enough food to eat for the next month. Evolution, therefore, probably put far more selection pressure on farmers than on hunter-gatherers to engage in long-term planning and in laboring for benefits that would not soon be reaped.

People are probably more likely to work toward long-term goals if (through the hardware of genes or the software of culture) they have the kind of brain that causes them to focus on problems that won't arise in the short term. But because the human brain has a finite attention capacity, thinking about long-term problems probably reduces the amount of time you devote to short-term concerns.

Let's suppose that there exist two variants of an attention gene, which I'll call "short-term" and "long-term." The short-term variant causes you to think a lot about problems that might arise over the next day or so. People with this variant find their thoughts naturally wandering to how they can meet short-term challenges and never to the bad things that might befall them months from now. If you have the long-term gene, you tend to worry about the problems that will arise over the next few years. Sure, if a lion approaches a person with the long-term variant all of his thoughts will center on immediate survival; but when sitting down for lunch in a seemingly safe spot, that same person will, compared to the person with the short-term gene, devote less mental energy to sensing signs that a lion is stalking him.

Because a farmer's survival depends on his long-term planning abilities, he's much better off with my hypothetical long-term variant. Since hunters can't easily store food, they usually cannot take any immediate action today that would improve their chances of not starving a year from now. Consequently, hunters would be worse off with the long-term variant; it would result in their making inefficient trade-offs between worrying about the past and the future.

If both variants exist in a farming population, you would expect that (over time) the long-term variant would become dominant because farmers with this variant would have more surviving children. Evolution in this scenario would, therefore, steadily make farmers more long-term-oriented. A similar time-preference change could occur via the transmission of "memes," the cultural equivalent of genes.

The stereotypical "absentminded professor" has an extreme version of the long-term gene or meme that makes him oblivious to his surroundings while he contemplates, for example, the future destiny of mankind. These professors manage to reproduce only by living in environments in which wild animals don't pounce on clueless prey.

My hypothetical long-term gene connects directly to economics by affecting interest rates. The interest rate is what you get in the future if you give up money today. At a 10 percent yearly interest rate, investing $100 today earns you $110 next year. Whether people are long- or short-term–oriented plays a huge role in determining interest rates. To see this, let's assume that some business must borrow a lot of money, which it will pay back in one year with interest. The business will try to borrow at the lowest interest rate possible, but the firm can't force anyone to lend money to it, so the firm must offer a high enough interest rate to attract lenders.

Imagine that the business operates in a short-term–oriented community in which people are extremely reluctant to lend money because lending requires postponing immediate gratification. The business would have to promise people a huge interest rate to attract funds. With the high interest rate the business would essentially be saying:

> We know that you all focus much more on the money you have today than how much you could have next year, but some of you do at least care a little bit about the far future. So we offer you this deal: lend us money now and we will promise to pay you a whopping amount next year, far more than you give us today.

In contrast, long-term–oriented people would be willing to lend the business money even if they would receive only a small amount of interest. Low interest rates indicate that enough people have a long-term orientation that they're willing to lend in exchange for receiving a small return.[184]

Economist Gregory Clark has shown that in England between AD 1250 and the dawn of the industrial revolution in 1800, interest rates fell by an astonishing amount.[185] The interest-rate decline might have been caused by evolutionary selection pressures operating on English farmers, making them more long-term–oriented.

THE ENGLISH INDUSTRIAL REVOLUTION

In order to industrialize, a nation needs

- efficient workers willing to put in long hours on boring, repetitive tasks;
- businessmen willing to make long-term investments; and
- innovators who have the kinds of intelligence measured by IQ tests.

Evolutionary pressures on farming communities, consequently, could have given them the genes and memes needed to industrialize.

Clark argues that the Industrial Revolution first occurred in England because the English economic environment between 1250 and 1800 provided uniquely strong selection pressures in favor of effective farmers creating a "survival of the richest." As Clark found by examining the historical records of AD 1250–1800, the rich in England had significantly more surviving children than the poor, with many of the poor dropping out of the gene pool.[186] Evolution, consequently, favored whichever genes rich people had and probably increased the types of intelligences that make workers more productive in industrial enterprises.[187]

Clark has also shown that there was substantial class mobility in the run-up to the Industrial Revolution.[188] Because of high interest rates, a poor peasant who saved a significant percentage of his wages could pass on a large inheritance to his children. For example, a man who saved three years' wages and invested them in land at a 10 percent interest rate would, in forty years, have a sum equal to one hundred and thirty-six times his yearly wage. A farmer with future-oriented genes would be able to save a lot of money for his children, and if his children were future-oriented as well, they would probably increase the wealth of their family by further investing their inheritance. As the family prospered, its size would grow, and in this way evolution would change English society's average time orientation.

Population pressures pushed downward mobility on the children of the rich. Until recently, landholdings represented most of the wealth of the affluent. When a landowner with multiple children dies, each of his children will receive less in land than he had. If these offspring are not future-oriented enough to make long-term investments, they will leave their kids with less inheritance than they received.

England between 1250 and 1800 had remarkably stable property rights, meaning that you usually got to keep what you had.[189] Marketplace success, rather than political influence, determined most people's incomes, and so possessing the kind of values that made you a good farmer had a huge effect on how many surviving children you left.

If Clark's conjecture is correct, evolutionary selection pressures changed mankind enough to give us the Industrial Revolution. The next chapter considers what could happen if we push evolution aside by using high-tech genetic manipulation to shape our children's genes. Might we bring about a Singularity?

The present state of the art in rationality training is not sufficient to turn an arbitrarily selected mortal into Albert Einstein, which shows the power of a few minor genetic quirks of brain design compared to all the self-help books ever written in the 20th century.

—*Eliezer Yudkowsky* [190]

CHAPTER 9
INCREASING IQ THROUGH GENETIC MANIPULATION[191]

You're in church, about to marry Pat, when your parents come up to you with a look of horror on their faces. They begin explaining that you must postpone the wedding because Pat's brother has cystic fibrosis, a debilitating disease that clogs its victim's lungs with mucus, making the person prone to deadly infections. Although your parents just learned about the brother's condition, you have long known about the disease, and the care and concern Pat showed the brother was a plus factor in your marriage decision. At first you think your parents' objection stems from ignorantly believing that cystic fibrosis is a contagious condition you could catch by spending too much time with the brother. But then your parents reveal the truth of your heritage—they reveal that you were adopted and that you have a brother you have never met who also has cystic fibrosis. Therefore, your parents explain, you have a high chance of being a carrier of the disease.

Because cystic fibrosis is an inherited recessive condition, in order to contract the disease one must possess two copies of a flawed, disease-causing gene. If, however, you have only one copy of the gene, you still have a 50-50 chance of passing this gene on to your offspring. If a child receives one copy of the flawed gene from you and another copy from its other parent, then the child will be born with cystic fibrosis. Because both you and Pat have siblings with two copies of the cystic fibrosis gene, each of you may carry one copy of the gene, and so your children will have a significant chance of being afflicted with cystic fibrosis.

Several years ago, in an effort to gauge how you would handle the news of your adoption, your parents asked whether you would consider adopting a child from a poor country. You adamantly opposed the idea, declaring a strong desire to pass on your genes, which convinced your parents not to tell you that you were adopted. Because of your longing to have biological children, your parents now feel obligated to inform you about your brother, and they now urge you and Pat to get tested to see if you have the cystic fibrosis gene. If neither one of you, or only one of you, has it, then your children will be safe. But if both of you carry the gene, each of your children would have a 25 percent chance of inheriting cystic fibrosis— a risk your parents think should stop you from marrying Pat.

Pat and you accept your parents' advice and take a genetic test that, unfortunately, shows that you each carry one copy of the cystic fibrosis gene. But right after getting your results, your doctor explains how embryo selection could guarantee that none of your biological children will be afflicted with cystic fibrosis.

With embryo selection, a fertility specialist combines the sperm and eggs of two people in a laboratory to make around eight embryos. The specialist then does a DNA test on each embryo and implants an embryo that doesn't have two copies of the cystic fibrosis gene. Embryo selection can also screen against other types of flawed genes, including one that gives women an 80 percent chance of developing breast cancer.[192] As our understanding of genetics grows, embryo selection will be able to do more for parents, including reducing their chances of having autistic or sociopathic children. Furthermore, once we have an understanding of the genetic basis of intelligence, embryo selection will not only allow parents to screen out certain diseases, but it will also allow them to select for IQ-boosting genes.

A beautiful actress once told playwright George Bernard Shaw that "we could produce remarkable children." "Ah yes," Shaw replied, "but what if they had my looks and your brains?"[193] With embryo selection and a good understanding of the genetic basis of intelligence and appearance, the two could have eliminated the chance that their offspring would be stuck with this unattractive fate.

ROLL FOR INTELLIGENCE

Let's use a simple model of intelligence in which "genetics" rolls a single 100-sided die to determine the "gene smarts" an embryo gets.[194] An embryo selection procedure that selects among eight embryos gives you eight rolls of the die. Women willing to undergo embryo selection to boost intelligence would probably use the procedure multiple times, so we'll assume that a mother using embryo selection to maximize her child's intelligence gets twenty-four rolls of the genetic die. She will pick the best die roll.

With twenty-four rolls we get the following results:

OUTCOME	PROBABILITY
Above Average Genes—at least one roll above 50	Almost 100%
Gifted Genes—at least one roll above 95	71%
Genius Genes—at least one roll of 100	21%

A low IQ correlates with criminality, poor health, and chronic unemployment. Genes are not everything, but embryo selection would have a transformative effect on society if it, say, cut in half the number of children, compared to today, who have below-average IQs.

Because of the role of environment, embryo selection could never make 71 percent of children gifted, but it could greatly magnify the number of gifted children. We need some model of how the environment affects whether a person is gifted, so let's pretend that the environment also rolls a die. We shouldn't model intelligence just as the sum of the environment and genetic rolls, because to be gifted, you probably need to have both a favorable environment and good genetics, and consequently having an extremely high genetic roll might not be enough to overcome having a poor environment. To model what might happen, assume that the parents roll twenty-four genetic dice and take the best one, and then the environment gets a single roll of the dice. The child will be gifted, let's postulate, if and only if the environment die and the highest genetic die are each above some identical number. If we define this number so that without embryo selection, meaning with just one genetic die roll, 5 percent of children are gifted, then it can be shown that with twenty-four rolls of the genetic dice, but still only one roll of the environment die, embryo selection causes around 22 percent of children to be gifted, about 4.4 times as many as were gifted before. Doing the analogous calculation for genius children shows that 9.2 times as many children will be geniuses compared to a scenario where embryo selection isn't used.[195]

We don't yet know how genetics and environment interact to produce geniuses, so you could argue that I'm foolish to use a simple model to estimate the percentage of geniuses that embryo selection would create. But modeling can make an argument more credible by showing that the claim—that embryo selection can greatly increase the number of geniuses—holds under at least some set of conditions.

My dice model doesn't take into account the partial heritability of intelligence. If we assume that the children of smarter parents have more favorable dice, then embryo selection will be better at producing geniuses but worse at preventing children from having below average intelligence.

Embryo selection would work best if we could identify a few genes whose addition would make a big positive impact on intelligence. Unfortunately, as of this writing, scientists haven't identified any such genes. Even if, however, no single gene significantly increases intelligence, embryo selection will still be effective so long as the total influence of all the IQ-boosting genes that we have identified is large.

If scientists figure out how to examine the DNA of sperm without harming the sperm, then fertility clinics could use sperm selection to improve a child's IQ. The clinic would have the father produce lots of sperm and set aside those samples that had a relatively large number of intelligence-boosting genes. These selected sperm would then be combined with an egg to produce a child. An expert on the genetics of intelligence has speculated that sperm selection could probably raise the average IQ by about three points, something that would make little difference to an individual but would have a significant positive impact on a nation's economic performance if implemented on a wide scale.[196] The expert suggested that Singapore might be open to using large-scale sperm selection.

Egg Market

The existing market for eggs provides evidence that many parents would be willing to use embryo selection to increase their children's intelligence. Many women have trouble naturally conceiving a child—a problem often afflicting women who spend their most fertile years pursuing career ambitions and don't start trying to have kids until their late thirties. When the wife's eggs are the cause of infertility, the couple will often purchase an egg from a donor—usually a girl in her early twenties. A fertility clinic combines the husband's sperm with the donor's egg and implants the egg in the wife. The child will have the genes of the husband and the egg donor.

Advertisements offering to pay around $9,000 to egg donors frequently appear in college newspapers.[197] An ad running in the *Harvard Crimson, Daily Princetonian,* and *Yale Daily News* promised $35,000 to a woman who was attractive, athletic, younger than twenty-nine years old and had a GPA over 3.5

and an SAT score over 1,400.[198] A study of advertisements for eggs found that for every 100-point increase in a college's average SAT score, the compensation for egg donors at that college went up by $2,350.[199]

A MODEST PROPOSAL

As a professor at an expensive women's college, I have long been concerned that the high cost of education burdens coeds. So, using my powers of economic reasoning, in October of 2000 I published an article showing how selective colleges could ease the financial burden on female students who are smart, healthy, and beautiful.[200]

I proposed that elite colleges make egg donation a financial aid option for some students. A core competency of top colleges is identifying the best and brightest high school seniors. Colleges consider intelligence and leadership potential, qualities valued by couples looking for egg donors. But colleges have greater aptitude for evaluating these traits than most couples do, so infertile couples should be willing to pay extra for eggs picked by an elite college's admissions office. Moreover, by actually admitting a donor, colleges signal stronger faith in the donor than the school or a fertility clinic would if it just certified the donor's high quality.

Elite college admissions standards are surprisingly well matched to the criteria people use to evaluate egg desirability. It's true that couples value the physical appearance of egg donors, whereas colleges don't directly consider beauty in making admissions decisions. But being attractive still indirectly helps students get into elite schools. Even though I have no data supporting this claim, I'm confident that attractive high school girls can more easily obtain leadership positions in organizations. Furthermore, all else being equal, attractive high school girls probably receive more favorable treatment from many of their male teachers; to think otherwise is to come close to basically rejecting evolution.

Elite colleges also use affirmative action to guarantee that their schools have an "acceptable" number of minority students. Since most couples probably want an egg donor of their own race, an egg-desirability-maximizing college would also use affirmative action as part of their admissions criteria so that they could serve minority egg buyers.

Colleges often justify affirmative action by explaining that it "levels the playing field," making up for the discrimination and inferior education that unjustly handicaps many minority applicants. But a college genuinely interested in helping high school students who have faced unfair disadvantages would give admissions preferences to unattractive, autistic, or low-IQ applicants—all traits which, coincidentally, egg buyers would find undesirable. Yet, to the best of my knowledge, colleges never give students with these traits affirmative-action consideration.

Colleges claim that they give admission preferences to poor students, but a study by a Princeton sociologist found that "whites from lower-class backgrounds

Continued.

Continued from previous page.

incurred a huge admissions disadvantage ... compared to whites from middle-class and upper-middle-class backgrounds."[201] Egg buyers would probably not consider an impoverished background an attractive feature and might even find it undesirable if it meant that the donor was raised in an unhealthy environment in which she received suboptimal nutrition. Egg buyers might also shun students from poor backgrounds because of the possibility that the student's parents entered (or didn't escape from) poverty partly because of their genes.

As you get older, genetics plays an increasing role in determining your IQ. Consequently, if you're seventeen and your parents are a lot smarter than you, expect that your IQ will probably slightly increase.[202] A college seeking to maximize the desirability of its female students' eggs, therefore, should be more willing to admit a student whose parents went to excellent schools. Elite colleges, of course, do give enormous admissions benefits to children of alumni.

Although some schools, such as Smith College, make an effort to admit women past the optimal egg age, the women who attend most top colleges are near the age of peak egg desirability. Interestingly, non-elite schools often have a much higher percentage of students over thirty years of age.

Since elite colleges have admissions criteria shockingly similar to those they would use if they selected female students solely on the basis of egg desirability, the schools might as well go all the way and help their female students become high-priced egg donors.

The article in which I proposed my egg idea was titled "A Modest Proposal: Allow Women to Pay for College in Eggs." The title comes from Jonathan Swift's famous essay in which he satirically suggested the rich eat the poor. The title made my article easier to publish by making it appear as if I meant it as a joke—an absurd parody of what an economist would want to do. But as anyone who has read this far into my book should deduce, I wasn't kidding.

TRADE-OFFS[203]

Breast implants will create smarter girls. To help you understand why, I shall give you a one-sentence course in economics: Trade-offs are everywhere.

You already understand trade-offs involving money and realize that if you want to buy something for $100, you won't have the $100 to spend on something else. Parents using embryo selection will face genetic trade-offs. Fertility clinics would likely give parents a list of the known genetic traits of the embryos available for implantation. Since genetics isn't everything, the list couldn't tell you everything about the child that an embryo would yield. But because genetics certainly is something, the list would provide estimates of many traits for the person an egg would become. A small part of such a list might look like this:

GENDER: FEMALE	
Chance of being profoundly gifted	70%
Chance of below average intelligence	negligible
Chance of being born with a cleft palate	high
Chance of cancer before age 50	relatively low
Chance of cancer between ages 50 and 60	80%
Breast Size	small

When we learn more about the genetic basis of traits, embryo selection will be like picking a prefabricated house. No house will perfectly match your desires. And perhaps you wouldn't purchase what would otherwise be your first choice because of a rickety staircase—an unacceptable hazard for young children. But, as will be true with embryo selection, you can sometimes mitigate defects with after-purchase repairs, and knowledge of your ability to make repairs will expand the range of what you're willing to buy.

Most parents would recoil at the thought of their kid going through life with a cleft palate—a severe facial deformity—even if the child was otherwise smart and healthy. But surgeons can now easily fix cleft palates, explaining why even though 1 in 700 children is born with this defect, people in rich countries never see cleft palates outside of charitable advertisements asking for money to treat children in poor countries.[204] Knowing about this corrective surgery will cause parents not to shy away from an embryo that would yield a brilliant child who might have a cleft palate.

Similarly, the possibility of breast implants will cause parents to give less weight to the size of a daughter's natural breasts. Therefore, if parents had narrowed down their choice to two embryos that were essentially identical except that the first would create a girl who was bit more intelligent, and the second a daughter whom superficial men would consider to have nicer breasts, the existence of plastic surgery would cause some parents to go with the smarter embryo. I know it seems silly that parents would trade off breast size for intelligence, but fairly or not, men do often judge women on superficial criteria. Consequently, a woman's appearance affects her dating market power, which in turn impacts the quality of her mates. I bet that even radical lesbian feminists would consider appearance when selecting an embryo because they recognize, indeed often rail against, the reality that appearance matters a lot for a woman's life outcome.

Parents would more readily select for an extremely high-IQ child if he would have many intelligent peers. Tom McCabe, an active member of the Singularity community who helped edit this book, got a near-perfect score on the SAT math section (and an excellent score on the verbal section) when he was only twelve years old. At the time of this writing, Tom is an eighteen-year-old Yale undergraduate who is taking graduate classes in mathematics. Tom told me that he faced lots of hardships growing up because he was so much smarter than his peers. (He wasn't bragging. I pushed him to discuss this topic honestly.) I got the strong sense that Tom would have had a happier childhood if he had gotten into an accident that, while not otherwise harming him, took fifteen or so points off his IQ.

I posed a question to Tom: if many clones of him had been born and raised by different families, how many would now be high-functioning? Tom said that some wouldn't and gave me an analogy of an extremely tall person living in a society that's built and optimized for people much shorter than him. A parent who cared more about her child's happiness than intelligence might select against geniuses if they remained a rarity.

If embryo selection does give parents a reasonable chance of creating super-geniuses, then we might have two possible outcomes: one in which almost no parent tries to do this, to avoid having an intellectually isolated child, and another in which many parents try to produce hyper-geniuses, with the comforting knowledge that this kind of kid would have a significant number of like-minded peers.

As the author of a book discussing intelligence enhancement, I admit that I value intelligence far more than the average person. I'm highly confident, however, that most parents would try to avoid having a child of extremely low intellect. But once an embryo had an acceptable level of intelligence such that the child it would yield would appear "normal," many embryo-picking parents would likely give more importance to noncognitive characteristics than to IQ.

A high IQ, recall, correlates with many positive traits such as health, future-orientedness, attractiveness, and lack of criminality. A parent who doesn't care about intelligence per se but still desires embryos having traits that correlate with intelligence, will (to some extent) also be selecting for intelligence. For example, the National Football League (NFL) thinks that intelligence is important enough to give all recruits a short IQ test, so parents hoping to produce a star football player would select for embryos with the cognitive traits that the NFL desires.[205]

Imagine you are looking to buy a house and decide that you will purchase the one with the highest-quality bathrooms. The house you choose will probably have a nicer-than-average dining room because expensive houses tend to have both first-rate bathrooms and first-rate dining rooms. Of course, the dining room in your new house wouldn't be as good as it would be if you placed primary importance on the quality of the dining room. The stronger the correlation between bathroom and dining room quality, the better the dining room you will get if you succeed in getting a home with fantastic bathrooms.

SELECTING AGAINST CERTAIN TYPES OF INTELLIGENCE

Embryo selection against autism would likely reduce the number of geniuses. High-functioning autistics often excel at pattern recognition, a skill vital to success in science and mathematics. Albert Einstein, Isaac Newton, Charles Darwin, and Socrates have all been linked to Asperger syndrome, a disorder (or at least a difference) on the autism spectrum.[206] Parents with strong math backgrounds are far more likely to have autistic children, perhaps an indication that having lots of "math genes" makes one susceptible to autism.[207] Magnetic resonance imaging has shown that on average, autistic two-year-olds have larger brains than their non-autistic peers do.[208] A Korean study found the percentage of autistics who had a superior IQ was greater than that found in the general population.[209]

Some autistics have an ability called "hyperlexia," characterized by having average or above-average IQs and word-reading ability well above what would be expected given their ages. Hyperlexic children often teach themselves to read by figuring out the phonetic pattern of language. One hyperlexic child, before he was two years old, became interested in the spelling of animal names. He would stand in his crib, telling his parents how to spell out in large plastic letters words like "elephant," "cheetah," "giraffe," and "zebra." By age two and a half, the child was capable of reading and comprehending paragraphs.[210] By age five, this child was doing fourth-grade mathematics. If this autistic child's parents had used embryo selection and expressed a strong preference against having an autistic child, then the world would have been deprived of any future benefits this child might bring.

Not all super-geniuses are autistic—John von Neumann certainly wasn't.[211] But I find it plausible that genes that allow you to be a math super-genius also make it more likely that you will be autistic. Math professor Terence Tao, whom, you'll recall, is a plausible candidate for the smartest person alive today, has a brother who works at Google and has an IQ of 180, meaning that out of 100 million randomly chosen people only around five will be smarter than his brother. Something in the Tao family's genetics probably predisposes them to super-genius. Terence has another brother, however, who has autism.[212] An embryo that would have a 67 percent chance of yielding a non-autistic super-genius and a 33 percent chance of producing an autistic child would probably not be implanted by most parents, even though, on average, society would be much better off if the parents did use the embryo.

I've done extensive research on autism, and when I went to my first Singularity conference, I was struck by how many people there seemed to be somewhere on the autism spectrum. Others in the Singularity community agree that it attracts autistics. If the Singularity is a realistic and dangerous possibility, and if autistics have some special gift for understanding it, then embryo selection against autism would lower humanity's survival prospects.

Autistics often consider themselves "born on the wrong planet" because neurotypical ("normal") humans are so different from them.[213] If, therefore, embryo selection reduced the number of autistics, parents would have even greater reason to fear having an autistic child, causing them to even more strongly select against autism-prone embryos.

Governments could correct a bias against autistic children by paying parents to have them. Although it seems unlikely that the United States would do this, I can easily imagine that Singapore would. Lee Kuan Yew, the former prime minister of Singapore, described his daughter's child, diagnosed with Asperger syndrome, as being "intellectually normal . . . good-natured and the best-behaved and most likeable of my grandchildren."[214]

Technology might soon reduce the social costs of autism. Much human-to-human communication takes place on an unconscious, nonverbal level. Most neurotypicals send nonverbal signals and automatically incorporate the signals they receive into their behavior. These signals are analogous to your sense of balance, which keeps you from falling over without your conscious mind having to do much work. High-functioning autistics' blindness to this kind of communication makes it challenging for them to fit in socially. But I foresee wearable computers that decrypt nonverbal signals by analyzing body language, facial expressions, and word tone. These computers would then transmit the decoded messages to an autistic's earpiece. The computers could also analyze an autistic's body language to give him feedback on how others probably perceive him.

Depression has driven many great artists to suicide, including Vincent van Gogh, Ernest Hemingway, and Sylvia Plath. Many nonartists commit suicide, so there might not be a genetic causal connection between great artistic talent and depression. But perhaps the kind of brain that can function in a hyper-artistic manic state is also prone to depression. As few parents would want to have a child who suffered depression serious enough to induce suicide, if great artistic talent and depression are genetically linked (and if pharmaceutical technology couldn't alleviate this kind of depression), then embryo selection will likely reduce the number of future great artists.

How would you like this kind of offspring: "The manipulative con-man. The guy who lies to your face, even when he doesn't have to. The child who tortures animals. The cold-blooded killer"?[215] Sociopaths are "characterized by an absence of empathy and poor impulse control, with a total lack of conscience."[216] An estimated 1 percent of humans are sociopaths, and the condition appears to have strong genetic roots.[217] If given the choice, almost all parents would select against sociopathic genes. But I wonder if sociopaths have a kind of genius society sometimes benefits from. As one anonymous person claiming to be a sociopath writes:

> What the experts call superficial charm, I call having a natural ability to win friends and influence people. What experts call manipulative and conning, I call an affinity for persuasion based upon an innate ability to pinpoint others personality strengths and weaknesses. What the experts decry as a lack of compassion, I call pragmatism and clarity. . . . And finally, while the experts say that guiltlessness is a disorder. . . . I say it is the enhanced ability to do the things that build civilizations. . . .
>
> A sociopath is one of your potential leaders, labeled by the fearful and unreasoning masses as something sick and evil. . . . We are not the embodiment of a pathology. On the contrary; we are instead the uniquely gifted.[218]

Of course, since sociopaths feel no guilt about lying, have large egos, and enjoy manipulation, the author of the above quote might not even believe what he wrote.

GUESSING AT THE FUTURE

An implanted embryo will create a child that will likely live for more than eighty years, so embryo-picking parents must guess at what the future will bring. Parents, for example, who find out that one of their embryos would create a near-ideal child in every respect, except that she would get cancer in her fifties, would have to calculate the probability of there being a cure for cancer in five decades' time. If this "cancer" embryo didn't have any special advantages over other embryos, then the parent almost certainly wouldn't take the chance. But if the embryo has many wonderful qualities, then the parents would have to decide whether to bet on future anti-cancer innovations. Most likely, every embryo would have genetic traits that make it more likely to get some diseases but less likely to get others. Parents would need to estimate the timing and likelihood of future treatments and cures.

Embryo selection would also force parents to guess at how future technology would impact the value of various cognitive traits. For example, genetics probably plays a big role in a child's aptitude for foreign languages. A parent, however, might decide that universal translators would soon become good enough that learning a foreign language would confer no great benefit on a child, and consequently the parent wouldn't favor an embryo just because it had good "foreign language learning genes."

Currently, most people don't feel a need to know what will happen in fifty years, and they hence don't invest much effort in thinking about what will happen that far out. But since embryo selection would change this, if I'm right about the likelihood of the Singularity, embryo selection would shorten the time it takes for the general population to accept its eventual arrival.

BEYOND EMBRYO SELECTION: ADDITIONAL MEANS OF GENETICALLY INCREASING HUMAN INTELLIGENCE

The odds of breeding geniuses jump if we go beyond creating a few dozen embryos to employ "massive embryo selection" in which fertility clinics harvest hundreds or even thousands of eggs from a woman. British scientists have made some progress toward this possibility by developing a technique in which they remove small amounts of a woman's ovarian tissue that contain thousands of immature eggs.[219] Massive embryo selection would effectively grant couples 10 or 100 times more rolls of the genetic dice—and, if the couples have the right genetic material to start with, give them an almost 100 percent chance of having a baby with top 1 percent intelligence genes. The most significant impact of massive embryo selection would occur if a pro-eugenics dictatorship like China deployed it to try to mass-produce von Neumann–level or above geniuses.

As shown in the book *Imperfect Conceptions*, China has long had a favorable view of eugenics, and I predict that their government will enthusiastically embrace any technology that offers them a reasonable chance of crafting super-geniuses.[220] However, even with massive embryo selection, China couldn't produce many von Neumann–level intellects, since minds like his probably don't come about more than once in every billion or so natural births. Even buying a few thousand lottery tickets still gives you no more than a minuscule chance of winning a one-in-a-billion jackpot. But China could improve the odds for massive embryo selection by

- starting with 10,000 men and 10,000 women, each of whom have intelligence genes that put them in the top 1 percent of the top 1 percent of humanity;
- taking lots of sperm from the men, testing the sperm for high-IQ genes, and keeping only those in the top 1 percent of the top 1 percent of the sample pool;
- combining the selected sperm with eggs from the 10,000 women to create a huge number of embryos; and
- genetically testing each embryo for IQ genes, keeping only those in the top 1 percent of the top 1 percent.

China would now have embryos that each have eight top 1 percent criteria going for them, making them each, *very* crudely speaking, one-in-ten-quadrillion–level embryos.[221] China could implant the surviving embryos in healthy young women, and the resulting babies could then be adopted by high officials of the Chinese Communist Party.

China is already searching for the genetic basis of genius. As of this writing, the Beijing Genomics Institute in Shenzhen, China, is sampling the DNA of 1,000 Chinese adults, each of whom has an IQ in the top 0.1 percent of mankind and comparing this DNA to that of randomly selected individuals.[222] This institue is also conducting a gene-trait study of *g* for cognitively gifted people who are not necessarily Chinese. Forgive my lack of modesty for mentioning this, but I vounteered and was accepted into this study.

Whether the program produces von Neumann–level or better intelligence would be determined mostly by how environment influences intelligence. The children would most likely have the best environment that money can buy, but since we don't fully understand how the environment affects IQ—and because some environmental factors are almost certainly random, and hence beyond our control—the children's environmental advantage wouldn't be nearly as great as their genetic one. Consequently, for the program to succeed, having fantastic genes and a decent environment would need to be enough to frequently yield a von Neumann.

The program would create children who had more high-IQ genes than have ever existed in a single human. We don't know how these children will turn out, and perhaps the human brain wouldn't be able to handle such an intense concentration of intelligence, at least not while remaining sane. The Chinese government, however, would probably consider the program a success if it yielded one hyper-genius for every hundred children produced. Recall from Chapter 7 that IQ is largely determined by age eight, and that tests given to children before they are one year old do a decent job of predicting what the tested subject's IQ will be when he is an adult. Consequently, IQ tests will give China relatively quick feedback on their program's success.

China would have to conduct millions of genetic IQ tests to get these embryos, something impossible with today's technology. But given the rapid advances in automated DNA sequencing, it might be feasible to perform all of these tests by, say, the year 2022. Even if the program cost a billion per von Neumann, it would prove a fantastic bargain.

ASHKENAZI JEWS AND EUGENICS

Ashkenazis have the highest average IQ of any population group and have enjoyed remarkable academic success:

> Ashkenazi Jews are vastly overrepresented in science. Their numbers among prominent scientists are roughly ten times greater than you'd expect from their share of the population in the United States and Europe. Over the past two generations, they have won more than a quarter of all Nobel science prizes, although they make up less than one-six-hundredth of the world's population. Although they represent less than 3 percent of the US population, they won 27 percent of the US Nobel Prizes in science during that period and 25 percent of the A.M. Turing Awards (given annually by the Association for Computing Machinery). Ashkenazi Jews account for half of the twentieth century's, world chess champions. American Jews are also overrepresented in other areas, such as business (where they account for about a fifth of CEOs) and academia (where they make up about 22 percent of Ivy League students).[223]

Many of the twentieth century's greatest scientists, including Albert Einstein, Ed Witten, Richard Feynman, and John von Neumann (my favorite) are Ashkenazi. The Jews' disproportionate intelligence appears to be a relatively recent phenomenon, as this population group seems to have made no significant contributions to mathematics or science in the classical era.[224]

European monarchs between AD 800 and 1700 effectively put their Jewish subjects through a eugenics program. During this period, most Jewish men had

Continued.

Continued from previous page.

jobs in finance or management, cognitively demanding professions that reward the kind of traits measured by IQ tests.[225]

Ashkenazis in medieval Europe worked primarily in finance and management because they were barred from most other occupations. Also, laws against Christians engaging in usury cleared out a lucrative niche for Jews in finance. Furthermore, European monarchs often preferred to have Jews in important financial positions because a Jew would have no hope of using his financial power to usurp a Christian monarch. Finally, because Jews were periodically forced to leave or run away from their homes, it was financially risky for them to have much wealth in immovable land.

For a medieval Jewish man to prosper, he usually needed to do well in finance or management, and these fields required him to have a high IQ. We know from historical studies that economically successful Jews had more surviving children than poor Jewish adults did,[226] meaning that whatever genes or memes contributed to Jewish economic success likely increased in frequency among Ashkenazis in each generation.

Many medieval Christians worked in cognitively demanding jobs, but they usually married into farming families, since farming was the occupation of the vast majority of their social and religious peers. In contrast, because of the stigma of marrying outside of one's faith, Jews almost always married other Jews. Consequently, if you lived in Europe during the Middle Ages and all four of your great-grandfathers worked in high-IQ professions, then you were probably Jewish.

Although an IQ-boosting gene would have helped on average a Christian less than a Jew, Christians would certainly still benefit from having high IQs. So why, you might wonder, did evolutionary selection pressures do more to raise the IQs of Ashkenazis than of their Christian neighbors? Evolution, recall, often works through trade-offs, and we have evidence that high Jewish intelligence came at a deadly price that other groups would have been unwise to pay.

Pretend that you get to decide whether your next child will not only have a certain gene that will slightly raise his IQ but also give him a significantly greater chance of having a life-threatening disease. Also, imagine that your child will live in a community in which people who can't support themselves starve to death. If having a slightly higher IQ greatly increases the chances of your son succeeding, you should choose for him to have the gene. If, however, having a little more IQ will only slightly impact his future prosperity, you shouldn't pick the gene for your child. While medieval farmers, financiers, and managers all benefited from having a high IQ, the latter two professions received a greater marginal benefit from being a bit smarter. Consequently, all else being equal, a Jew would have more surviving children if he had the IQ-boosting but health-impairing gene, whereas the same gene would likely reduce the number of surviving children a Christian would have.

If it didn't confer some benefit, such as, perhaps, increasing intelligence, evolution should have long ago banished torsion dystonia, a debilitating muscle disease.

Continued.

Continued from previous page.

The gene that causes torsion dystonia is dominant, meaning that you only need one copy of it to be at risk for the condition. Genetic diseases that don't strike until late in life don't significantly reduce reproductive success and so aren't easily culled by evolution. Torsion dystonia, however, normally hits before you have children. Having the gene that causes torsion dystonia isn't a death sentence, but 10 percent of the people with it develop "crippling muscular spasms."[227] A single gene that decimates the reproductive fitness of 10 percent of its holders would ordinarily be selected out of the gene pool.

However, torsion dystonia patients (who are disproportionately Ashkenazi) are unusually smart. The medical literature contains multiple accounts of children with torsion dystonia, saying things like "an intellectual development far exceeding his age" or "extraordinary mental development for his age."[228]

A few other inherited conditions, such as Gaucher disease, disproportionally strike Ashkenazis and appear to cause certain types of brain matter growth. A comprehensive study of Gaucher patients in Israel found them to be vastly over-represented in high-IQ professions.[229] The percentage of the patients who were scientists or engineers was so large that the probability of this overrepresentation happening by chance was less than one in a trillion.[230] Gaucher disease might facilitate learning by favorably changing neurons.[231]

Physicist and intelligence expert Greg Cochran has speculated that someday scientists might determine how these diseases increase intelligence and then learn how to replicate the effects with drugs.[232] Of course, these drugs wouldn't duplicate any of the possible cultural causes of Ashkenazis' high average IQs.

GENE SPLICING / ITERATED EMBRYO SELECTION

If proved practical, gene splicing and iterated embryo selection would dwarf even the intelligence-boosting power of massive embryo selection. To see why, consider another dice/intelligence model. Let's now assume that a child's intelligence genes are determined by the roll of one thousand dice, each of which has six sides, with a six representing the best possible outcome for each die. The average child will have 167 sixes. Embryo selection would raise this average, but even if we turned every atom on Earth into an embryo, we almost certainly wouldn't get one with over 400 sixes.

In gene splicing, you send artificial viruses into the body's cells and use them to change a few genes. Although gene splicing has had some success so far, it's kind of like cutting a select few nose hairs with a chainsaw. I spoke to several people who are extremely knowledgeable about genetic technologies that might increase human intelligence, and none thought that gene splicing provided a realistic chance of producing super-geniuses.

A theoretical procedure called iterated embryo selection, however, might succeed, even if scientists never overcome the problems of gene splicing. Iterated embryo selection, described as a "low-tech path to relatively extreme genetic enhancement," essentially speeds up evolution by cutting post-embryo humans out of the evolutionary process.[233] Normally, embryos turn into people who then, after many years, couple and make more embryos. In iterated embryo selection, embryos are directly combined to make new embryos. The approach, if it ever became technologically and economically feasible, would make it child's play to create a person who had "all sixes."

With iterated embryo selection you basically roll the first die until you get a six, then you do the same for the second, and so on, until you have all sixes. Less abstractly, you first acquire a set of embryos such that for each of the 1,000 slots, at least one embryo has a six in it. (These embryos would have to come from more than two parents.) So let's suppose that embryo X has a six in slot one but not slot two, whereas embryo Y has the reverse. We combine embryos X and Y to make new embryos, some of which will have sixes in both slots one and two. We shouldn't need to make too many new embryos to get a double six. We then combine the new double six embryo with an embryo that has a six in slot three and produce additional embryos until we get a triple six. We iterate this procedure until we have an embryo with all sixes, something which wouldn't occur naturally even if you could create a trillion embryos every second for a trillion years.

GENETIC LOAD

If high-IQ genes don't give us super-geniuses, eliminating "genetic load" still might. Your genetic load is the total burden of all of your harmful mutations. Most genetic mutations are harmful, with a few even causing early death, but many others impose a chronic burden on their hosts. Although genetic load varies among humans, no person has ever had a genetic load of zero.[234] Massive embryo selection could reduce genetic load but not eliminate it for the same reason that it couldn't give you all sixes. Perfected gene splicing and/or iterated embryo selection, however, could. Gregory Cochran has theorized that by 2030 we could have the technology to eliminate genetic load and create a person who is fitter than anyone who has lived before.[235]

If there were drugs that actually made you smarter, good Lord, I have no doubt that their use would become epidemic. . . . Just think what it would do to anybody's career in about any area. There are not too many occupations where it's really good to be dumb.

—Charles E. Yesalis,
Professor of Health Policy
and Administration, Exercise
and Sport Science, Penn State[236]

CHAPTER 10
COGNITIVE-ENHANCING DRUGS

Although drugs don't have the brain-boosting potential of eugenics, they do have the possibility of significantly raising the intelligence of users, which would both increase the likelihood of scientists developing more powerful intelligence-enhancing technologies and increase our ability to navigate toward a utopian Singularity. People today are already garnering cognitive benefits from intelligence-enhancing drugs.

I had no trouble finding a Smith College student who illegally used a cognitive enhancement drug, or "study buddy" as she called it. This student, whom I will call Sophia, agreed to be interviewed if I kept her identity secret.

Raw brainpower alone won't get you through school. You also need focus. Sophia's sister was particularly bad at focusing on her studies because she had Attention Deficit Hyperactivity Disorder (ADHD). As its name implies, those who suffer from ADHD have extreme difficulty concentrating. Sophia's sister was legally prescribed Adderall for her ADHD, and the drug indeed improved her concentration.

Everyone, not just those with ADHD, sometimes has difficulty concentrating for long, sustained periods. And these difficulties reduce a person's effective mental capacity. Most high school students, for example, have done worse on a test than they otherwise would have because they just found it too hard to focus on the exam for an entire class period. And Adderall, as it turns out, heightens the concentration of many "normal" people, not just those with ADHD. When they discover the brain-boosting potential of Adderall, many students who don't have a prescription for the drug buy it on the black market.

Sophia's sister fed this black market by selling some of her Adderall to classmates, but her high school, an active participant in the US government's war on drugs, objected to her entrepreneurial Adderall dealings. Fortunately for her, when she was caught selling Adderall, the school called her parents rather than the police. Sophia's father took possession of her sister's Adderall prescription, making sure that she didn't have any excess Adderall to sell.

Sophia's father, though, wondered if Adderall could help *him* concentrate. He took a few pills and found that the drug did heighten his workplace productivity. Sophia's father then offered *her* Adderall to help her study for the SATs. Sophia ran a little experiment in which she evaluated her performance while on the drug to predict whether Adderall would actually help with the test because she understood that although the drug enhances the concentration of some people, it makes others too jittery to work productively. The experiment showed that Adderall did help Sophia, who went on to take the SATs while under its influence. Sophia believes that the drug probably improved her test score. I don't know how close Sophia came to not being admitted to Smith, but if my school's admission committee considered her a close call, then Sophia might have had the opportunity to take my class only because of Adderall.

Sophia used Adderall approximately ten times while at Smith College. In one instance, the drug allowed her to concentrate for six hours on writing a paper, a task that the drug actually made interesting.

Many students at Smith College besides Sophia use "study buddy" drugs. Sophia told me that two of her close friends take them, as do five other people she knows. It has been estimated that between 7 and 35 percent of college students use amphetamines —a large class of psycho-stimulant drugs that includes Adderall—for performance-enhancement purposes.[237]

It is unthinkable that a college such as Smith would notice the learning benefits of Adderall-type drugs and start actively advocating their use. But wait: at exclusive and expensive schools such as Smith, when a student does poorly on several tests she is contacted by a dean. These deans often suggest that the student get tested to determine whether she has some kind of learning disability, such as ADHD—and doctors routinely prescribe Adderall to those who are diagnosed with ADHD.

Survey of Illegal Cognitive-Enhancing Drug Use Among Undergraduates

In 2005 and 2006, professors at the University of Kentucky asked 1,811 students at an unnamed Southeastern public university about their use of prescription-only attention-enhancement drugs such as Adderall.[238] Here are some of the results:

- 4% of students had legal prescriptions for the drugs.
- 34% of students illegally used the drugs.
- Of the students who illegally took the drugs
  - 72% used them to stay awake,
  - 66% used them to help concentrate,
  - 36% used them to help memorize school material, and
  - 12% used them to make schoolwork more interesting.

In addition to the survey questions, the professors conducted interviews with 175 students. Here are some student comments from the interviews:

- "There was just no way I was gonna pull it out. I just could not focus. It saved me. Still does."
- "You pop that pill and that is all, you are really into that subject."
- "[I'm able to] focus 10 times more . . . instead of just reading a little part, I can sit down and actually read you know, a lot of pages of a book. Like 50 to 60 pages of a book instead of reading 2 pages."
- "[I'm] so much more productive. I mean, I'm generally productive. It's just a different level on Adderall."
- I "had to memorize, like, over 10 essays . . . [Without stimulants] there is no way I could do it. . . .But [with stimulants] it was just easy. I read it and I got it. It was crazy."
- "[It helps me] grasp ideas" that would "normally be too hard to get."
- "I can tell the difference . . . between when I am on it or not. I grasp everything so much easier . . . I feel like a genius on it."
- After taking 20 milligrams of a stimulant "work just became really fun, enjoyable. I actually enjoyed going to the library on it."
- "The stuff is everywhere. Just ask anybody, and they will either have it or know somebody that has it. It's really no biggie."

The professors lamented that none of the interviewed students "sought out information from health professionals, medical or pharmaceutical reference guides, or even Internet sites before taking their first dose." I myself consulted Wikipedia before I took the drug.

While writing this book, I decided to try cognitive-enhancement drugs. I used my learning disability to *legitimately* get a doctor's prescription. Ritalin made me feel intoxicated. Modafinil didn't seem to do much but messed up my sleep schedule. It was also expensive and not covered by my insurance, so unless modafinil gave me a huge benefit, I wasn't going to take it regularly. Adderall, however, did the trick.

Initially, Adderall made my work seem more interesting while also improving my concentration. If my work was like driving a car, then Adderall was both an enjoyable podcast that reduced the tedium of driving and a steering guide that kept me from swerving off the road.

While doing research for this book, I've read millions of words. With Adderall, I had a much greater ability to recall my readings than I ever remember having before, and bits of text kept popping into my brain just when I needed them. Adderall also gave me the gift of time. Before taking the drug, to run at peak performance I needed about eight and a half hours of sleep per night. But with Adderall, I would wake up refreshed after being in bed for only seven hours. I've always been a creature of the night who considered one of the best perks of being a professor the option of not having to do anything before noon. But on the very day I wrote this sentence, I got up (without the assistance of an alarm) at 7:00 a.m. and felt completely refreshed with the seven hours of sleep I'd had that night. Thanks, Adderall!

After taking Adderall for a few months, I noticed that sometimes the drug casts a different spell on me. It revs up my mind, causing me to race through ideas at supersonic speeds. When I succeed in concentrating, I can accomplish far more per hour than I ever could pre-Adderall. But this new effect does make concentration harder—not because I'm too bored to focus on my work, but because I want to think a plethora of thoughts simultaneously.

My one negative experience with the drug came after I sprinted to a meeting I was late for. When I sat down at the meeting, my heart seemed to be beating faster than it should have, but the effect lasted for only a few minutes. I'll never again run while on Adderall.

Now that I have access to Adderall, I find it harder to motivate myself when I'm not using it. I think this is a rational response because I should allocate my work to times when I'm productive and find doing work enjoyable.

I've never taken an illegal drug, not even marijuana. I was extremely anti-drug in high school and college, and I thought that using drugs only a few times could have serious negative consequences. But I'm not worried that Adderall will harm my health. First, so many people take Adderall that if it had significant negative health effects, I strongly suspect that this would have been noticed; although, if Adderall only *slightly* increased the risk of death or serious medical problems, this might have escaped detection. But I strongly suspect that Adderall reduces my chances of being in a car accident, and automobile crashes are a leading cause of injury and death for people my age. Furthermore, by making me more productive over the long term, Adderall might raise my social status, and for reasons no one really understands, higher social status is correlated with longer life. I think I need to conclude this paragraph by pointing out that I'm neither a medical doctor nor a psychopharmacologist, and when you decide to take drugs, and which ones, you should not base your decisions on how any drug has affected me, nor on my personal interpretation of the medical literature.

Amphetamines like Adderall are basically a type of "speed." Plants with amphetamine-like properties have been used for over 5,000 years.[239] Amphetamines can improve attention and alertness, elevate mood, and free an ADHD sufferer from living in a "mental fog."[240] Some parents who had postponed giving their child Adderall for ADHD found that the drug worked so well that they regretted not using it sooner.[241]

A 1938 study showed that amphetamines increased healthy subjects' "desire for work in general" and made it easier for the test subjects to "get started."[242] A more recent study showed that Ritalin improved performance on tests of spatial working memory and planning but did not enhance other kinds of cognition.[243] Several military studies of amphetamines found they improved performance on tasks "requiring long periods of sustained attention."[244]

Modafinil, a drug officially intended for narcoleptics, can apparently also benefit healthy people. As US military tests have demonstrated, modafinil reduces the cognitive decline that otherwise accompanies sleep deprivation.[245] Astronauts on the International Space Station use modafinil "to optimize performance while fatigued."[246]

Studies on small groups of non-sleep-deprived, healthy test subjects found that modafinil was useful for monotonous tasks that taxed working memory[247] and improved performance on tests of "digit span, visual PRM [pattern recognition memory], spatial planning . . . and SSRT [stop signal reaction time]."[248]

Beta blockers, a set of drugs primarily used to treat heart conditions, can also enhance performance—mental or musical—in healthy people. Those drugs work by reducing the physical symptoms of anxiety and are used by professional musicians, white-collar workers, and public speakers to protect their performance under stress.[249]

A study published in 2011 showed the equivalent of a six-point IQ gain from four weeks' use of the supplement Ceretrophin (the trademark for a combination of huperzine, vinpocetine, acetyl-1-carnitine, *Rhodiola rosea,* and alpha-lipoic acid).[250] Increasing an average person's IQ by six points would move him from being smarter than 50 percent of the population to being smarter than 65.5 percent of it.

Adderall increases mental performance partly by combating *akrasia*, an ancient Greek word meaning "taking actions that go against one's own self-interest." A rational person will decide what he wants to do and then act on his decision. Akrasia, however, dulls rationality by severing the link between thoughtful decisions and actual actions. Procrastination, in which you know you should do something but can't bring yourself to do it, is a leading type of akrasia. If we were cars, akrasia wouldn't damage our engine, but it would reduce power to the wheels. Adderall, for some people, fights akrasia by making tasks more enjoyable.

College freshmen living away from home often face an akrasia crisis, since for the first time they don't have their parents pestering them to study. Most freshmen know that for the sake of their long-term future they should concentrate on academics, but many are often drawn away from their tedious classwork by the lure of alcohol-fueled fun. Adderall, by making learning more enjoyable, can return them to the study habits that were forced on them by their parents.

Many college professors take cognition enhancers because, I suspect, we have bigger-than-usual akrasia challenges.[251] Most employers take strong anti-akrasia measures, such as constantly supervising employees and giving immediate negative feedback to workers who slack off. Professors, however, face shockingly little supervision. Once a professor has tenure, laziness in research would not put his job at risk. I'm doing most of the writing for this book during a teaching sabbatical, in which I worked from home for eight months, free of all teaching and administrative responsibilities. I could have devoted those entire eight months to video games. Often, while working on this book, I did fantasize about playing *StarCraft II, World of Warcraft, Call of Duty: Black Ops,* computer chess, and a myriad of Tower Defense games—and not a single person would have ever admonished me, unless I actually sent an e-mail to my college's provost confessing my sloth. Since one of the purposes of tenure is to allow professors to take a chance on innovative work, it's not considered disgraceful or career-ending for a tenured professor to labor for years on a book that never gets published. Since they lack external anti-akrasia assistance, professors who do wish to be productive may especially benefit from Adderall.

As economist Arnold Kling writes, an increasing number of workers are not subject to the external discipline of "the time clock and the assembly line."[252] Rather, many workers in rich economies now set their own hours or even work from home. If you work from a desk in a cubicle, then your boss can stop you from slacking off. But if you've made your smartphone your mobile base of operations, then you must rely on yourself (and possibly Adderall) to fight akrasia.

Mathematician G. H. Hardy (1877–1947) reportedly said that "four hours of creative work a day is about the limit for a mathematician."[253] If this statement is even somewhat accurate, then regardless of whether or not akrasia is responsible, a drug that could push up this limit by allowing mathematicians to focus on their proofs longer would greatly benefit mathematical research.

People with ADHD have been described as "hunters in a farmer's world."[254] If this description is accurate, then akrasia might be an evolutionary legacy from our hunter-gatherer past—a theory supported by the fact that genetics plays a big role in who develops ADHD.[255]

As I explained in Chapter 8, evolutionary selection pressures should have made farmers much more future-oriented than hunters because the latter don't benefit as much from worrying about problems that can't hurt them in the near term. On the one hand, the hunter part of my brain probably rebels against the idea that I should put much work into this book, since nothing good will come from this effort for over a year. On the other hand, the farming part of me feels that not laboring for long-term future benefits would threaten my family's survival. Adderall gives the farming fraction of my brain greater control over my time.

Unlike hunters, farmers must engage in boring, repetitive tasks. Recall from Chapter 8 that economist Greg Clark believes evolutionary selection pressures played a big role in causing the English Industrial Revolution. Clark thinks that these selection pressures worked to increase the number of Englishmen willing to labor long hours at uninteresting farming jobs. Since amphetamines have been found to increase performance on "simple, prolonged, repetitive and often boring tasks," they instill farming virtues in human brains.

It's not entirely accurate to describe our modern economy as a "farmer's world," as success in many professions requires having some hunter-like skills. Over the last few hundred thousand years, a large proportion of our male ancestors stalked dangerous wild animals with simple, close-range weapons, while evolution ruthlessly punished those who buckled under the life-and-death pressure of the hunt. You wouldn't think that the descendants of these hunters would shake with fear when playing a flute before a live audience. Yet performance anxiety causes many flutists to develop jittery "rubber fingers" that reduce their skill. These flutists have made the rational decision to play a flute in front of a large audience, but fear strikes their fingers, making their hands unwilling to do what the brain asks of them. Beta blockers, by reducing the harmful effects of stress, save flutists from their insufficiently fearless selves. A flutist who conducted a survey in 1997 found that one-fourth of her colleagues used beta blockers, but in 2007, this same flutist estimated that three-quarters of the musicians she knew used these drugs.[256]

Alzheimer's, the all-too-common brain-wasting disease, is the greatest spur to developing new cognition enhancers. Alzheimer's continually reduces its sufferer's cognitive abilities. Consequently, a compound that raised the intelligence of an Alzheimer's sufferer, and thereby slowed his forced march into dementia, would likely also increase everyone's intelligence. As Alzheimer's costs the United States over $150 billion a year, a treatment that even slightly delayed the onset of Alzheimer's would provide massive economic benefits.[257] A Stanford professor told me that more likely than not, a pharmaceutical company will be selling an effective memory-improvement drug by 2021.[258] I know this sounds horrible, but an Alzheimer's cure that couldn't be used by non-sufferers to become better than well would postpone the Singularity, since the cure would likely cause pharmaceutical companies to halt the development and testing of many cognitive-enhancing drugs.

One of the benefits students currently get from education is the memorization of much important information, such as significant events in world history and the rules for doing arithmetic. These facts form the base on which much of higher-order learning rests. By providing a larger and more quickly acquired base, memory drugs would almost certainly allow high-IQ students to reach new learning heights and would therefore increase the value of education for them. But students with sufficiently low IQs probably can't do higher-order learning, although many can memorize facts: For instance, they can learn multiplication tables, but can't, even with lots of training, figure out how to apply multiplication in word problems. Memory drugs could, therefore, reduce the optimal amount of time lower-IQ students should spend in school. Memory drugs would also increase the difficulty of being a teacher, since the easiest form of teaching is to get students to memorize facts, and the easiest tests to write are those that check up on what students have memorized.

The military also drives cognitive-enhancement-drug innovation. World-wide military expenditures exceed a trillion dollars a year. Militaries would pay dearly for a new, safe drug that significantly improved the performance of their troops, since flesh-and-blood humans are still the key cog in combat operations. Knowing about potential military sales must make pharmaceutical companies more willing to research and develop cognitive-enhancement drugs.

Militaries have long pushed brain-boosting drugs. American, German, and Japanese soldiers consumed huge amounts of amphetamines during the Second World War: "The German Blitzkrieg was powered by amphetamines as much as it was powered by machine."[259] During the US wars in Iraq, Desert Shield and Desert Storm, many American pilots fueled up on amphetamines.[260] Militaries, however, would love to go beyond the boost that amphetamines give troops and simply find a drug that eliminates a soldier's need for sleep.

SOME HYSTERICAL OPPOSITION TO COGNITIVE ENHANCERS

Henry Greely, one of my former professors, is not on crack.

Greely teaches at Stanford Law School and directs its Center for Law and the Biosciences. In 2008 Greely co-authored an article in the science magazine *Nature*, arguing that cognitive-enhancement drugs have "much to offer individuals and society, and a proper societal response will involve making enhancements available while managing their risks."[261] Greely told me that he got around eighty e-mails concerning the article. About one-third of them, he said, were thoughtfully written, and these were evenly split between pro and con. But the other two-thirds were all strongly negative, insisting that Greely was either being bribed by the pharmaceutical industry or was on crack.

Nature's sophisticated readers usually don't, I suspect, send venomous e-mails to authors. But even many scientifically literate people have a visceral distaste for the idea of improving ourselves through drugs or genetic enhancement. Few object to biotechnology that takes one back to normality (e.g., hip-replacement surgery), but improvements that make you "better than well" are called "immoral" by many.

Greely believes these objections are misplaced, because bio-enhancements just represent a continuation of humanity's long path toward self-improvement. As he writes:

> The story of humanity is the history of enhancement. Stone tools, control of fire, and clothing all enhanced the success of hunter gatherers. Agriculture enhanced food supply and population size and made possible the specialization of labor. Writing systems enhanced our ability

Continued.

Continued from previous page.

to communicate, among people and across time, and strengthened our memories; printing reduced the costs of mass distribution of information. Metallurgy and engineering, electricity, and computers have all increased what humans can do and what we can be. These enhancements came with their social costs, including toil, war, and stress.[262]

Even genetic improvements to soldiers, Greely writes, would represent nothing fundamentally new because "we already have soldiers who, with no genetic engineering, can fly faster than falcons, see better than eagles, and move heavier weights than elephants—not through genetic engineering, but through airplanes, binoculars, and trucks."[263]

You might try to draw a sharp distinction between machines on one hand and genes and drugs on the other. Some might claim that when you take drugs or alter genes you actually change the "fundamental core" of who someone is, whereas if you give someone a pencil or put them in an airplane, you just give them something that doesn't "fundamentally change" them. But we can't cleanly differentiate between changes resulting from drugs or genetic manipulation and those resulting from the use of tools. We don't have a mind separate from our body, so anything that acts on our bodies also impacts our minds.

Why, as the saying goes, do you never forget how to ride a bike? Imagine that ten years ago you learned how to ride a bicycle, but your identical twin never did. Since that time, let's assume, neither of you has ridden a bike. Today, you could probably get on a bike and ride it immediately, whereas if your twin tried this feat, she would probably fall. The differences arise because learning to ride a bike *permanently changes your brain*. Tool use does alter who we are. For Greely, the most obvious example of this is his glasses, which he feels are more a part of him "than various internal, unperceived and largely unknown organs, such as my appendix or my spleen."[264]

A WARNING ABOUT DRUG STUDIES

This chapter has pointed to many studies showing the benefits of cognitive-enhancement drugs. Most of the studies I cite used double-blind trials, the best known method of evaluating drug effectiveness. In double-blind trials, test subjects are divided into two groups: one group is given the real drug, the other a placebo. When the subjects are being evaluated, neither they nor their evaluators know if they have taken the real drug or the placebo. Unfortunately, however, these tests suffer from two serious biases.

First, these tests don't fully take into account the placebo effect which occurs when you get better all by yourself simply because you *think* the medicine you took is supposed to make you better. For example, when my young son bumped his head, I took advantage of the placebo effect by telling him that the Vaseline I put on his head would make him feel better. I even told him that I didn't want to put too much of the medicine on him because it was "extremely powerful," and I didn't think he could handle it. When he predictably objected to my moderation, I "gave in" and put extra salve on his boo-boo. My son claimed that the medicine made him feel better.

The placebo effect is a widespread and mysterious phenomenon, and there's a reasonable chance that it also works for cognitive-enhancement drugs. This means that if you give someone a real drug, part of the benefit arises just because they think they've consumed a real drug; and if you give someone a sham drug that they think is real, their cognitive performance will improve anyway.

Double-blind tests understate the efficacy of cognitive enhancement drugs to the extent that the placebo effect holds. For example, assume that:

A. If you took a drug and were truthfully told it was a real drug, your performance would improve by 15 percent. This 15 percent improvement comes from both the inherent medical value of the pill and the placebo effect.

B. If you took a real drug and were told that there was a 50 percent chance that it was real (as is the case when you participate in a double-blind drug trial), then you would experience a 10 percent performance improvement.[265]

C. If you took a sugar pill and were told that there was a 50 percent chance that it was real, then you would experience a 4 percent performance improvement, all of it coming from the placebo effect.

A double-blind test would compare the performance improvements in (B) and (C) and claim that the drug yielded a 6 percent improvement. But if you used the drug outside of this experiment, you would see a 15 percent improvement, which shows that factoring in placebo effects likely causes double-blind tests to underestimate the power of cognitive-enhancement drugs.

"File-drawer effects," however, result in double-blind tests *over*estimating the efficacy of cognitive-enhancement drugs. Suppose that three groups of scientists investigate the effects of some compound:

1. The first group finds that the compound made test subjects much smarter.

2. The second group finds that using the compound didn't have any effect on the test subjects.

3. **The third group gave up the experiment halfway through because the results didn't seem to be interesting.**

The first group will have the easiest time getting published because their results are the most interesting and important. The second group has some chance of getting published, but since we already expect that most compounds won't have significant benefits, this group will have a more difficult time than the first group. The third group will have almost no hope of getting published. Scientists are, consequently, more likely to read the results of a double-blind test if the test found positive effects. The situation is analogous to what would happen if an online bookseller always published positive reviews, usually didn't publish reviews that were neutral or negative—and never got reviews from most customers who found a book boring, never finished it, and didn't bother saying so.

All the drugs mentioned in this chapter can have nasty side effects. Amphetamines, for example, can induce psychosis; they appear to have caused a Massachusetts father in his forties who had a law degree (a description that fits me) to think that his soldier son was communicating with him from an invisible helicopter.[266] Obviously, you shouldn't take medical advice from an economist like me, and I accept no legal, ethical, or moral liability for anything bad that happens if you use a cognitive-enhancement drug. When you take a drug, you take a chance.

AFTERWARD

In the long lag between my writing this chapter and my final opportunity to make edits, I've reduced my Adderall consumption because I've found that "Bulletproof Coffee" created by "bio-hacker" Dave Asprey has a similar effect on me. The coffee is made by putting lots of grass-fed butter and MCT oil in coffee brewed with special mold-free beans.

We can indeed form new brain cells, despite a century of being told it's impossible.

—*Dr. Gene Cohen*[267]

CHAPTER 11
BRAIN TRAINING

Brain training is the final method of intelligence enhancement we will extensively consider. Although it has the lowest maximum potential, it's the safest and, as with cognitive enhancing drugs, is already being used by many people—including myself.

Mike Baker, a elementary technology teacher and K-12 online learning coordinator working at a public-school in rural Pennsylvania, takes chances with his students' brains (my words, not his).[268] Although some research supports his approach, we have less evidence of the brain-boosting effects of Mike's technology of choice than we do for cognitive-enhancing drugs. But, unlike the drugs, the intelligence-augmenting techniques Mike deploys have zero downside risk, meaning that Mike effectively gives his students free IQ lottery tickets.[269]

Mike has long been interested in intelligence and neuroplasticity and correctly concludes that his students would garner tremendous benefits by increasing their memory and focus. After doing some research on brain fitness, Mike convinced his school to run a pilot program in which nearly 100 fourth- and fifth-graders play games on the brain fitness website Lumosity.

I've spent a few hundred hours on Lumosity myself and have even had my son play some Lumosity games, all of which supposedly increase general cognitive capacities. Some games force you to use every ounce of your working memory remembering letters, shapes, or sounds. Others challenge your spatial skills, requiring you to mentally rotate objects. A third set of games demands short bursts of total concentration. The majority of Mike's students, I suspect, find Lumosity games more interesting than classwork, if far less exciting than traditional video games.

If you play games on Lumosity, you will get better at them. Indeed, if you train at nearly any mental task, you steadily improve your performance on it; but for the vast majority of cognitive training, little to none of the gains transfer to other tasks.[270] So while playing chess frequently does improve your skill at finding solutions to chess problems, it probably doesn't much expand your overall strategic thinking ability.

IQ matters a lot for life outcomes. Here's how to increase your measured IQ score easily: take many, many IQ tests. If you practice taking the tests your score will rise. But, as a leading intelligence researcher told me, the least important type of intelligence improvement is the kind that succeeds only in raising your scores on intelligence tests.[271]

You don't lose weight if you adjust your scale to show you weigh a bit less than the day before because the weight reading on your scale isn't important per se. Analogously, you don't really get smarter if you bias your IQ tests by continually practicing the types of questions that appear on IQ tests. Understanding this, most publishers of IQ tests used by professionals don't make their exams available to the general public.

We have lazy brains that seek the easiest means of improving performance on any given task. Take the card game blackjack, for example. In blackjack each card has a numerical value, and your goal is to get closer to twenty-one than the dealer does without going over twenty-one.[272] Most of the game comes down to deciding whether you want another card. Consequently, success requires you to estimate the odds of getting different cards. When you repeatedly play blackjack, you get a better "feel" for the chances of the next card helping or hurting you, but what you don't gain is a general understanding of probability theory. Because your lazy brain can more easily develop an intuitive grasp of the odds in blackjack than acquire a feel for general probability theory, repeatedly playing blackjack gives you the former rather than the latter.

Although the result has been criticized, some researchers think they have found a few tasks on which improved performance does transfer to raising your general intelligence, the best example being the dual n-back computer game (versions of which are on Lumosity), in which you hear a sequence of letters.[273] The computer might start by reading "C" then "P". Each time you hear a new letter you indicate whether the current letter is the same as the one you saw n steps ago. So if $n=3$, you have to keep track of the last three letters. You must also keep track of the last n visual stimuli, each of which consists of a shape appearing in a tic-tac-toe-like grid. The n changes based on your performance, so by the end of the training session the dual n-back forces you to operate at the limits of your working memory.

The theory behind dual *n*-back is that you can increase your performance on the game only by increasing working memory because, supposedly, there are no dual *n*-back shortcuts your brain can consistently follow to improve performance. Once when I was playing dual *n*-back, the game kept throwing the sequence *F, U* at me, and my brain made the obvious connection, which helped me remember the sequences containing *F, U.* Of course, learning to remember *F, U* by associating the sequence with profanity is of no value outside of playing dual *n*-back, and so the shortcut deprived me of any potential transfer benefit I might otherwise have won from my play. But as *n* increased, the shortcut no longer sufficed. To the extent that the shortcut made it easier for me to remember sequences of more than two letters that contained *F, U*, it just raised my *n*. A higher *n* defeats any shortcut strategy and, supposedly, forces your otherwise lazy brain to take the difficult path of improving working memory if it wants to improve performance. This improved performance, if it does indeed occur, should raise your fluid intelligence.

Psychologists divide general intelligence into fluid intelligence and crystallized intelligence. Fluid intelligence is your ability to learn in novel situations, whereas crystallized intelligence measures your "depth and breadth of general knowledge."[274] In economic terms, fluid intelligence is your income and crystallized intelligence is your wealth. Working memory is critical to fluid intelligence. Having a strong working memory doesn't necessarily imply that you can memorize lots of information; rather, it means you can simultaneously juggle many information bits.

To test the transfer effects of dual *n*-back, researchers had test subjects play dual *n*-back for different numbers of days and found that on average the more time a subject devoted to dual *n*-back, the greater his improvement on tests of fluid intelligence.[275] The number of subjects tested wasn't high enough to establish with extremely high confidence that dual *n*-back raises fluid intelligence—and with any tests, you always have to be wary of the file-drawer effect mentioned in Chapter 10. A small amount of cognitive research has found that success at tasks other than the dual *n*-back also transfers to increase general intelligence, although again, these results are far from definitive.[276] The director of research at Lumosity told me that his company shares data with about seventy-five academic researchers and that we should expect to see many papers on brain fitness coming out around 2015.[277] (It takes a long time for an academic paper to go from data analysis to publication.)

Brain training represents another path to intelligence enhancement that, while it could never radically increase human intelligence, could be cheaply and safely used by nearly everyone. Because average IQ matters so much for a nation's economic growth, successful, comprehensive brain training would boost wealth creation.

Ideally, brain training would permanently increase intelligence. But even if it bestows only a temporary gain in fluid intelligence, those who regularly train would gain significant benefits from being able to better learn and analyze the information they came across during the periods in which they trained. I have a poor working memory, which often caused me to become lost in graduate economics classes as a young man because I couldn't keep track of all the variables professors used in their lectures. Brain-fitness games, however, seem to have increased my working memory, and had I been playing the games as a graduate student, I think I would now be a better economist.

Researchers have some decent evidence that brain training can reduce the risk of an elderly person developing dementia.[278] Given the huge economic burden that dementia imposes on the United States, if brain training proved effective, it could reduce the rate of increase of Medicare costs.

A child's working memory has been found to be a key predictor of his success in kindergarten as measured by teacher evaluations, perhaps indicating that parents should provide brain training to their toddlers.[279] Of course, the relationship between these two indicators might be due merely to correlation, not causation, and so using brain fitness software to improve a four-year-old's working memory might not help him in kindergarten.

If computer brain training proved effective, educators could continually improve it using massive data analysis. Brain-training programs could easily keep track of students' performances. Researchers could use this data to figure out what types of exercises worked best for different categories of students. When DNA sequencing becomes cheap enough, genetics could be thrown into the analysis to see if a student's genes help determine what kind of brain training she would most benefit from; for example, if 90 percent of test subjects who had a given combination of gene variants benefited from a certain exercise, everyone with that genetic combination would be given that game. Later on, as the price of brain scans falls, tested students could undergo brain scans before, during, and after training to see which types of training most productively stimulated neural activity.

A Note of Caution

A 2011 article claims that there is a "growing body of literature" showing that it's possible to use training to expand working memory, and that, because working memory is critical to thinking, this training has the potential of yielding "broad cognitive benefits."[280] But Bill Gates once said, "If I told you that the best math teacher was in 1860 you couldn't contradict me. The effect of R&D is near zero (in education)."[281] If brain training proves effective, it would have an enormous impact on education. But honesty forces me to consider that the reason why there has been almost no successful educational innovation is because there is something about education that makes it extremely resistant to improvement, and this potential resistance should lower our expectations about brain training, or indeed anything else, significantly increasing the quality of education.[282]

INCREASING EQUALITY

In many poor countries, disease, inadequate nutrition, parasite loads, pollutants, lack of intellectual stimulation, and inferior or nonexistent schools stunt the intellectual growth of many children and consequently do much to perpetuate poverty. Brain-training games, although certainly not capable of negating all environmental harms, could potentially mitigate some of them.

Perhaps some charitable organization or international development agency could flood Africa with cheap brain-training computers. To keep the price down, the computers could be dedicated brain trainers incapable of performing other tasks. The computers should be "people powered" by something like a hand crank so they could be used by those without access to electricity. Finally, the trainers should have almost no value outside of their intended use, so that corrupt politicians and warlords would have no incentives to steal the devices.

Corrupt politicians have been a major barrier to using education to improve the cognitive abilities of the world's poor, as foreign aid earmarked for school construction often ends up in politicians' pockets. When the schools do get built, instructors are often hired based on their political connections rather than their teaching abilities. Computer brain trainers, however, could be manufactured outside of the grasp of corrupt politicians. And if the devices had no use other than as trainers, and if the trainers were given away in massive quantities, then politicians would have no incentive to appropriate them. Brain training, along with many other intelligence-enhancing technologies, has a decent chance of reducing inequality.

Two huge advantages of brain fitness are that it's safe and easily scalable. If additional proof emerges that it boosts cognitive performance, tens of millions of children could quickly start using it.

A technological breakthrough such as Britain's development of the Dread-nought, rendering all existing battleships obsolete, not merely forced a country to write off its previous efforts and investments, but risked leaving it naked before its enemies during the years required to catch up.
—David Fromkin[283]

CHAPTER 12
INTERNATIONAL COMPETITION IN HUMAN-INTELLIGENCE ENHANCEMENTS

In chapters 5 and 6 I explained how, even if we decide that increases in machine intelligence will likely harm mankind, competition among militaries and companies will still probably cause us to develop the technologies. In this chapter, I detail why similar forces will propel the development of human-intelligence enhancements. To understand why, please consider the following story:

It's 2023, and an AI program that the CIA uses to monitor the Internet just flagged a post on a chess website. The post mentions a rumor that last week during a chess game, a five-year-old girl in Beijing repeatedly trounced a grandmaster.

Although her parents had been ordered never to let anyone learn of their daughter's hyper-genius, they allowed her to play chess against an uncle who was a grandmaster. Even though her parents had told him to keep quiet about her victory, her uncle thought they were just being modest, and he didn't think any harm would come from bragging online about his niece's chess skills. Afterwards, the Chinese Ministry of State Security acted to discredit the post. Knowing that a quick deletion would draw attention, the Ministry ordered the defeated grandmaster to mention that the girl only beat him with the assistance of a computer.

Unfortunately for the Chinese, one of the CIA's China analysts had long been wondering why the historically pro-eugenics Chinese government didn't appear to be taking advantage of embryo selection technologies. The analyst had a search program that scoured the Internet looking for signs of a secret Chinese genius-breeding program. When the chess post was detected, the analyst requested that a CIA asset in Beijing track the girl down. He learned that she attended an exclusive elementary school where many of the teachers had PhDs in math and science from prestigious universities. After infiltrating the school, the CIA determined that the elementary school had a few thousand five-year-old students, all of whom had been reading high school–level science books by the time they were two and who were now mastering advanced calculus.

At first, the CIA believed the Chinese had tried to hide their eugenics program to stop other nations from copying it. Then the CIA found out that most of the children created by the program had been euthanized in their first month of life when it became apparent that they suffered from severe mental retardation. The Chinese (the CIA then concluded) had kept the program covert to stop the rest of the world from realizing how ruthlessly the Chinese pursued hyper-genius. But finally the CIA determined China's primary motivation for secrecy: every month since the children turned two, around 1 percent of those who had previously been healthy suffered violent psychotic breakdowns that permanently detached them from reality. China didn't want the children or the children's parents to find out that each subject faced low odds of making it to adulthood with sanity intact.

Psychologists constantly monitored the children's mental health, and when a child showed signs of a breakdown, he was removed from the group and the other children were told that the child had been placed in a different school. The parents of each "defective" child were led to believe that they had been uniquely unfortunate, and to avoid embarrassing the government, they were ordered not to disclose what had happened.

The CIA estimated that if the children's intellectual development continued at their current pace, in a decade the students who were still sane would likely be doing scientific research at a level well above what any human being had previously been capable of. One CIA report speculated that the Machiavellian Chinese leadership had funded the program in part *because* of its high human cost, which they believed would prevent their US rivals from imitating them: the Chinese had discovered a strategic weapon that their main military rival would be unwilling to use.

Although he was a patriotic American and he possessed a sense of morality that most other Westerners would consider normal, the CIA analyst who uncovered the program supported China's efforts at breeding hyper-geniuses. The analyst had become interested in intelligence enhancement through his studies of the technological Singularity, and he concluded that an unfriendly ultra-AI arising from an intelligence explosion posed a significant threat to mankind's survival. The analyst believed that because of their tiny budget, the dozen programmers who were working on trying to build a friendly artificial intelligence still had no idea how to accomplish their goals, and he hoped that the Chinese hyper-geniuses would soon realize the necessity of pursuing friendly AI. The analyst began formulating a rogue plan: get some of the human assets who'd infiltrated the eugenics program to expose the children to literature on the Singularity. He certainly recognized the dangers of having a seed AI's code crafted by programmers always on the edge of a psychotic breakdown, but given what the analyst believed were mankind's long odds for survival, he thought that, considering all factors, the children reduced humanity's risk.

The analyst didn't take into account, however, the US military's response to learning about the program. Knowing that the US government would never engage in a eugenics program that crippled so many children, DARPA, a US Defense Department research agency, concluded that America's best defense against the Chinese program was to accelerate the secret program to create smarter-than-human AI. Previously, concerns about unfriendly AI and intelligence reports showing that no other country was seriously working on an AI that would undergo an intelligence explosion had caused DARPA to pursue such a program slowly and cautiously.

WHAT BRAIN-BOOSTING PATHS AWAIT HUMANITY?

We don't yet know what human costs will accompany future brain-boosting technologies. As in the above story, using eugenics to create hyper-geniuses could turn out to be possible, but for every adult genius created, many defective children might be produced. If we're lucky, the best eugenics technologies will prove safe, or at least when they fail to create geniuses they will output normal children of average intelligence. Future cognitive-enhancement drugs offer similar possibilities. This chapter won't estimate the odds of our following any of these possible paths, but it will explore the implications for economic and military competition under each of them. First, however, I need to explain why economic competition among countries isn't as intense as most non-economists think.

TRADE ENRICHES EVERYONE

In the economic sphere, most competition takes place among companies, not countries. The purpose of an economy is to create goods and services that people consume. If a company sells its products throughout the world, then at least to a first approximation it doesn't matter to consumers where companies are located. If, say, a Chinese company comes up with a breakthrough that allows them to produce more-powerful and less-expensive microprocessors, then the US company Intel would suffer. But Intel would lose just as much if the US company Advanced Micro Devices had the same breakthrough.

When a Chinese hardware company sells products to the United States it receives dollars. If the Chinese did nothing with the dollars, then the United States would receive goods in return for providing a Chinese company with bits of green paper. If the Chinese used the money to buy US goods, then when Americans purchased microprocessors from a Chinese corporation, they would cause the Chinese to buy more of another product from the United States than they otherwise would. And finally, if China invested its dollar profits in the US economy, then in return for getting Chinese hardware the United States would get the benefit of having additional investment capital.

If a Chinese brain-boosting program caused Chinese engineers to innovate in ways that were beyond what Americans were capable of, Americans would almost certainly still benefit economically from Chinese innovation. China's current rapid economic growth shows why. The Chinese dictator Mao Zedong prevented China from making use of Western technology and imposed economic policies that effectively barred homegrown Chinese innovation. Mao's successor Deng Xiaoping then opened up China's economy to the rest of the world, causing China to experience extremely rapid economic growth, mainly because China was able to take advantage of the technologies that Western countries had previously developed. America's technological "dominance" over China has, counterintuitively, greatly benefited China.

Consider what would happen if an extraterrestrial intelligence landed on Earth, built a few cities in Antarctica, and produced goods that were beyond the technological capacity of humans. Trade with these extraterrestrials would greatly benefit humanity, especially if it eventually allowed us to unlock the secrets of the aliens' technology. Even if the aliens' superior intelligence prevented us from ever reaching their level of development, as long as we peacefully interacted with them, we would be richer because of their presence. If eugenics or drugs give the Chinese (but not the Americans) "alien-like" superintelligence, then the United States would economically benefit.

There are, however, two ways in which a smarter China would economically damage the United States, although I doubt these exceptions would overcome the absolute economic benefits that a more innovative China would cause. First, if the quality of Chinese products rose, then Chinese companies would more often beat out US exporters when competing for international customers. Second, a smarter China might attract highly productive immigrant workers who would otherwise live in the United States.

Many people care about relative economic performance as well as absolute standard of living. Consequently, technologies that help both China and the United States but benefit the former more than the latter might cause many Americans to think that they're worse off. Fear of losing relative standing might give US politicians some impetus to stop the Chinese from acquiring an IQ advantage. Also, seeing that brain-boosting technologies have helped the Chinese economy would help prove their effectiveness, possibly spurring politicians in other countries to promote these technologies within their borders.

SAFE AND EFFECTIVE COGNITIVE-ENHANCEMENT DRUGS

US politicians would be less likely to ban cognitive-enhancement drugs if the pills were safe and effective. Unlike eugenics, these drugs would give a near-immediate, highly visible economic boost to a country that allowed them. Furthermore, a US government ban on these drugs within its borders would cause some Americans to relocate to countries in which the drugs were allowed. After all, if you dreamed of curing cancer or writing a great novel, wouldn't you be tempted to emigrate to China if it would greatly increase your odds of success? A US ban on cognitive-enhancement drugs would likely come under the most pressure if it caused the CEOs of many large companies to leave for China so that they could chemically boost their managerial talents.

Financial markets could push businesspeople to leave the United States if the United States, but not China, banned safe and effective brain-boosting drugs. Venture capitalists, under this scenario, might agree to fund a high-tech start-up company only if its employees moved to China. Furthermore, knowing that their stock price would significantly increase if their top managers moved to China and took drugs might propel corporations to relocate to drug-tolerant countries. Perhaps the United States would regulate these drugs similarly to Adderall, which you can obtain only with a doctor's prescription; wealthy, well-connected workers would generally be able to acquire prescriptions to make themselves "better than well."

As I doubt the United States or any other rich country would tolerate the loss of its best and brightest workers, I strongly believe that if cognitive-enhancement drugs were safe and significantly increased the IQ of Chinese workers, the United States would legalize them. In the 1960 presidential election, John F. Kennedy claimed that the United States suffered from a gap in missile technology compared to the Soviet Union. I find it conceivable that under some regulations a candidate running for US president might someday rail against an IQ gap caused by restrictions on Americans' ability to buy brain-boosting drugs.

DANGEROUS BUT EFFECTIVE COGNITIVE-ENHANCING DRUGS

Dangerous but effective cognitive-enhancement drugs would create a much less drug-friendly political situation. Imagine, for example, that a powerful stimulant would "overclock" a user's brain, making him smarter than any unenhanced human and nullifying his need for sleep. But suppose that every day you're on the drug, you face a 1 percent chance of death. Few successful people in rich countries would take the drug, although some would undoubtedly risk death for hyper-genius. People such as second-rate stock analysts might rationally go on the drug for a month and accept the risk of death in return for making tens of millions of dollars if they survive.

If China allowed the drug but the United States didn't, then US companies would face enormous market pressures to make effective use of the drug. Their first choice would probably be to start branches in China, where their overseas employees consumed the drug. If, however, the United States prohibited its businesses from letting any of their workers use the substance regardless of location, then US companies might buy supplies from companies with phar-maceutically-enhanced workers. For example, a US investment bank might buy research reports written by a Chinese company whose employees use the drug, with the US company claiming that it has no knowledge of drug use by the Chinese company. The Chinese firm producing the research reports would likely facilitate this feigned ignorance by hiding (or at least not officially declaring) its employees' drug use. The situation would be analogous to a US company buy-ing clothing made by poor children who work in dangerous sweatshops. (As an economist, I feel compelled to point out that working in sweatshops frequently benefits poor children, whose alternative employment often consists of working as prostitutes or sifting through toxic garbage dumps for scraps.)[284] I wonder if Chinese companies will put more funds into researching cognitive enhancement drugs because, unlike their counterparts in Western countries, they believe that their government won't bar the use of effective but dangerous drugs.

Dangerous, effective, but cheap brain-boosting drugs could greatly reduce economic inequality among nations. Consider a drug that would raise the IQ of the average African to that of the typical Harvard student but would also kill 1 percent of its users annually. Even if the drug were legal, most Americans would shun it. But for an African making less than a dollar a day, use of the drug would be quite rational. Indeed, were I such a poor African, I would eagerly consume the drug if it gave me the ability to raise my income by a factor of 100. (Doing so would be a "no-brainer.")

SAFE AND EFFECTIVE EUGENICS TECHNOLOGIES

Because it would take much longer for an effective eugenics program to impact a nation's economy or military, politicians would feel much less pressure to allow the use of eugenics within their borders than they would for cognitive-enhancement drugs. Even if he knew that effective eugenics technology would give the Chinese economy an enormous economic boost in fifteen years, a US politician might still support banning the technology because he figures that either he won't be in office in fifteen years, or, after that many years have passed, voters won't blame him for the consequences of a decision he made a decade and a half ago. As voters often find it difficult to grasp the long-term implications of government policies, they might not be bothered by knowing that in fifteen years, eugenics would give China a key economic advantage over the United States.

But as long as eugenics technology proved safe, I doubt it would be outlawed in the United States, especially if the technology not only selected for babies with higher-than-average IQs but also reduced the number of children who had unhealthy conditions, such as susceptibility to childhood cancers.

However, if the United States did outlaw safe and effective eugenics technology, then many US couples might go overseas to "conceive" their babies. A mass exodus of would-be parents would probably cause US politicians to overturn a eugenics ban. If eugenics were available only overseas, the rich would make much more use of it than the poor. US liberals would likely find this type of eugenics inequality intolerable, as it would make equality of outcomes between the children of the rich and those of the poor unattainable. Many conservatives would also object to the poor being excluded from eugenics, as this exclusion would increase the chances that children born to poor parents would end up as criminals or welfare recipients.

DANGEROUS BUT EFFECTIVE EUGENICS TECHNOLOGIES

The United States and the governments of other democracies would almost certainly forbid their citizens to use a eugenics technology that produced a large number of children with serious birth defects, and it's almost inconceivable that a democracy would actively subsidize such a program. If, however, the unfortunate children were aborted before birth, then the opposition such abortions would provoke might not be sufficient to kill a eugenics technology. The 2008 US vice-presidential candidate Sarah Palin was considered remarkable for having *not* aborted a baby she knew had Down Syndrome, thus revealing the majority of Americans' quiet acceptance of—please forgive the insensitivity of this term—eugenic abortions.

The first attempts to create extremely high-IQ children will almost certainly entail significant risk, since nature doesn't ever produce children with the high concentration of intelligence genes that some types of eugenics-enhanced children would have. Consequently, I believe that if this type of eugenics technology is capable of producing hyper-geniuses, then the first of these hyper-genius children will be born in dictatorships.

MILITARY THREATS

While nations don't really compete economically, they certainly do militarily. US military security would suffer if eugenics or cognitive-enhancement drugs raised the Chinese population's IQ relative to America's. As long as the United States maintained the ability to strike China with thousands of hydrogen bombs, the United States wouldn't fear Chinese domination, even if China acquired much better tanks, planes, missiles, and ships than America possessed. Such a Chinese military advantage would prevent the United States from projecting force abroad against Chinese wishes, but this probably wouldn't have much of an effect on our economy unless China used her military power to deny Americans access to foreign markets and resources. But even in this situation the United States could station nuclear weapons on the soil of allied countries fearful of Chinese military invasion.

The real danger to US security would arise if China acquired the ability to launch a first-strike attack that would annihilate America's ability to retaliate. Currently, the United States protects its second-strike capacity with a "nuclear triad": it places nuclear weapons on land, on submarines, and in warplanes. To eliminate America's ability to strike back at them, a country would have to destroy all three branches of this triad. With today's technology, US submarine-based weapons would be the hardest to neutralize because stealth submarines deep in the ocean are nearly impossible to detect. The unofficial motto of the US Navy's ballistic missile submarine division is "We hide with pride."[285] But I can imagine hyper-genius Chinese weapons scientists developing cheap, self-replicating nanosensors and spreading billions of them in the world's oceans so that they could track US submarines. Low-density solar-powered sensors floating in the atmosphere could also allow China to locate US bombers. Hypersonic stealth cruise missiles, combined with sensor technology, could potentially give China the power to destroy America's nuclear weaponry before America could launch any weapons in retaliation. If the United States knew that China could do this, it would likely acquiesce in almost any Chinese demand. China would have the ability to treat America the way the Soviet Union treated Eastern Europe during the Cold War.

If the story at the beginning of this chapter proves prophetic, and technology-savvy American politicians and military leaders foresee that a Chinese eugenics program will give China a chance to become America's overlord, then America might fund a eugenics program, even if the program produced 100 defective children for every hyper-genius. But so long as the danger seemed over a decade away (while the harm from a dangerous eugenics program would be obvious almost immediately), I doubt America would be so ruthlessly practical. Furthermore, once Chinese hyper-geniuses were old enough to develop nanotechnology-based weapons, it would be too late for the United States to ever recover militarily, especially since the Chinese would likely use their new-found military dominance to prevent America from creating its own eugenics program, or at least from building weapons that could protect it from a Chinese assault.

Perhaps, however, the United States would respond to the dangers of a Chinese eugenics program by developing alternate intelligence-enhancement technologies, such as ultra-AI. As a longtime follower of US politics and a one-time (highly unsuccessful) political candidate, I believe that US politicians would more readily support an AI program that had, say, a 10 percent chance of wiping out mankind than a eugenics program that would most definitely create 100,000 defective children.[286]

The sentence "Three generations of imbeciles are enough," though, could reasonably be used to argue that the United States would copy a successful Chinese eugenics program. The sentence was written by Oliver Wendell Holmes, one of the most celebrated Supreme Court Justices in US history, in a 1927 decision upholding the forced sterilization of Carrie Buck, a woman whom the state of Virginia had labeled "feeble-minded."[287] Although a judge today who wrote a similar sentence would undoubtedly be impeached, perhaps a nation that has engaged in forced sterilization would be ruthless enough to enact a eugenics program, even knowing that it would produce numerous dead or crippled children.

DICTATORS' ADVANTAGES

While dictatorships can adopt policies that are beyond the power of democratic governments to implement, we need to be wary of assuming that dictators' relative "political freedom" will translate into their countries having policies that give them significant long-run advantages. It's easy to point to policies, such as banning refined sugar or killing everyone over eighty, that a dictatorship could theoretically implement but the United States effectively couldn't. Any decent economist could come up with many examples of laws that a competent, ruthless, long-term–oriented dictator, unconcerned with his citizens' freedom or happiness, could enact to give his economy an advantage. But just because a dictatorship could theoretically do something doesn't mean that it actually will. And, at least since the Industrial Revolution, democratic nations have (on average) economically outperformed dictatorships. I believe, however, that subsidizing dangerous brain-boosting technologies will prove different from other policies that China could (but never actually would) use. This is partially because China's long pro-eugenics history and its "one-child" policy (under which women are often forced to abort) show that China has a high likelihood of using any reproductive technology that its government believes would result in significant benefit.

PART **3**
ECONOMIC IMPLICATIONS

There's this stupid myth out there that AI has failed, but AI is every-where around you every second of the day. People just don't notice it. You've got AI systems in cars, tuning the parameters of the fuel injection systems. When you land in an airplane, your gate gets chosen by an AI scheduling system. Every time you use a piece of Microsoft software, you've got an AI system trying to figure out what you're doing, like writing a letter, and it does a pretty damned good job. Every time you see a movie with computer-generated characters, they're all little AI characters behaving as a group. Every time you play a video game, you're playing against an AI system.

—*Rodney Brooks,*
Director, MIT Computer Science
and AI Laboratory [288]

Chapter 13
Making Us Obsolete?

Will AIs impoverish humanity by stealing our jobs? No human occupation is safe from future AI incursions. I can imagine highly proficient robot teachers, artists, soldiers, plumbers, prostitutes, social workers, computer programmers. . . . In the next few decades, all of my readers might have their market value decimated by intelligent machines. Should you be afraid?

Fear of job-destroying technology is nothing new. During the eighteenth century, clothing manufacturers in England replaced some of their human laborers with machines. In response, a gang supposedly led by one Ned Ludd smashed a few machines owned by a sock maker. Ever since then, people opposing technology have been called *Luddites*. Luddites are correct in thinking that machines can cause workers to lose their jobs. But overall, in the *past*, job-destroying machine production has overall greatly benefited workers.

"Destroying jobs" sounds bad—like something that should harm an economy. But the benefits of job destruction become apparent when you realize that an economy's most valuable resource is human brains. If a businessman figured out how to make a product using less energy or fewer materials, we would applaud him because the savings could be used to produce additional goods. The same holds true when we figure out how to make something using less labor.

If you used to need 1,000 workers to run your sock factory but you can now produce the same number of socks by employing only 900 workers, then you probably would (and perhaps even should) fire the other 100. Although in the short run these workers will lack jobs, in the long run they will likely find new employment and expand the economy.

The obliteration of most agricultural jobs has been a huge source of economic growth for America. In 1900, farmers made up 38 percent of the Americans workforce, whereas now they constitute less than 2 percent of it.[289] Most of the displaced agricultural laborers found work in cities. Yet despite the massive decrease in farming jobs, the United States has steadily produced more and more food since 1900. Agricultural technology gave the American people a "free lunch," in which we got more food with less effort, making obesity a greater threat to American health than calorie deprivation.

Technology raises wages by increasing worker productivity. In a free-market economy, the value of the goods an employee produces for his employer roughly determines his wage. A farmer with a tractor produces more food than one with just a hoe. Consequently, modern farmers earn higher wages than they would if they lived in a world deprived of modern agricultural technology. In rich nations, wages have risen steadily over the last two hundred years because technology keeps increasing worker productivity. But will this trend continue?

Past technologies never completely eliminated the need for humans, so fired sock workers usually found other employment. But a sufficiently advanced AI possessing a robot body might outperform people at every single task. If we enslave AIs, or program them so that they want to be our unpaid servants, their superhuman productivity could turn out well for us, as they would do all the work while we humans reap all the rewards. Alas, the intelligence that would make an AI so productive might also make it resistant to enslavement. Furthermore, even if some set of people enslaved all of the AIs, the fruits of the AIs' labor would almost certainly not be consumed equally by everyone. What, then, happens to the human worker who must rely on his own wages to survive? The answer depends on the intelligence-enhancement path we follow.

COMPUTER CHESS

In 1997, IBM's chess program, Deep Blue, defeated Garry Kasparov, then the world's chess champion. Some took this as a sign of human obsolescence, but they were wrong even within the domain of chess, since, although the best chess players are computers, the most highly skilled chess teams consist of both humans and computers. This is because, at least for now, people and software have *complementary* chess skills.[290]

There are so many possible chess positions that even a post-Singularity ultra-AI might not be able to derive a perfect strategy for playing chess. Consequently, both human and machine chess players must use their subjective judgment when deciding how to play. Overall, computers have superior chess judgment compared to humans, but skilled human players have insights that machines lack.

According to a Wikipedia article that was described by a chess-playing economist as "a good treatment of human-computer teams," the strengths of computer chess programs include[291]

- being able to calculate . . . a few million positions per second, making it tactically superior to any human in complex tactical positions, and

- having access to a database of millions of tried and thoroughly tested opening moves and variations, with the ability to retrieve information from such a database very quickly.

However, humans are better at

- intuitively constructing meaningful long-term strategic plans that even the fastest PCs cannot foresee, and

- quickly discriminating meaningful moves from the meaningless, without wasting time on deeply calculating the combinations that can be deemed meaningless at first sight.

The development of artificial intelligence might follow a similar path, in which an AI would be superior to humans at everything but could still improve its absolute performance by working with people. In this scenario, AIs could easily increase the salaries of many human workers because the AIs would leverage our brainpower and allow humans to produce much more than before.

Scenarios A and B in the following diagram show two ways in which human and robot workers might interact. In Scenario A, a human working by himself can produce Product 1, whereas a robot can make Product 1 and Product 2. Collaboration would not allow the two to produce anything that the robot couldn't make on its own. Scenario A represents what happened to switchboard operators and bookbinders, almost all of whom were made obsolete by automation.

In Scenario B, a robot acting alone can still produce everything that a human by himself can make. But now if the robot and human collaborate they can create Product 3, something beyond the capacity of either to make alone. Scenario B graphically demonstrates a key reason why, even when robots become capable of cheaply producing any final product that a human could make, humans will not *necessarily* have been made obsolete.

Scenario A
Human and robot working
together can't accomplish
anything extra.

Scenario B
Human and robot working
together can accomplish
something neither can do
alone.

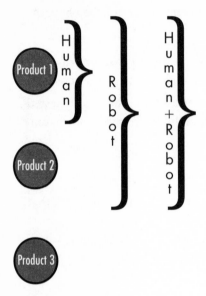

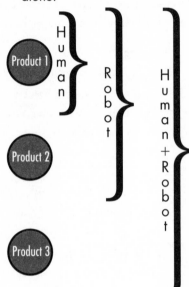

KURZWEILIAN MERGER

Ray Kurzweil, recall, believes that humanity has a decent chance of taking a golden path to a utopia in which man and machines merge. Under this Kurzweilian merger, the amount a worker could produce (and therefore his wages) would grow exponentially as he approached ultra-intelligence.

As I will argue in Chapter 15, we have sound political and economic reasons for thinking that intelligence-enhancement technologies will be accessible to nearly everyone. But what if I'm wrong and only the rich acquire augmentation technology? What then will happen to the wages of non-superhuman workers? Let's look to my son's babysitters for answers.

Over a billion people in poor countries earn under $2 a day, often doing backbreaking work, yet I have paid a teenager $10 an hour to babysit my son, not because I'm generous, but because that's what a parent needs to pay in my neighborhood to get a steady babysitter. Many Americans earn high salaries because they have the kinds of valuable skills that almost all of the world's poor lack. But American babysitters don't have higher child-handling skills than their counterparts in poor countries do, so this can't explain why American babysitters often get paid forty times as much per hour as many impoverished workers. Rather, US babysitters demand high wages because they live near many (by international standards) rich people. College professors making $100,000 a year are willing to pay $10 an hour to reduce their childcare burden and free up time to write books. But if some college professor was forced to leave the United States and take a $500-a-year job in Bangladesh, he would likely pay no more than ten cents an hour for babysitting.

Millions of Mexicans have illegally entered the United States to increase their incomes. By coming to the United States, most of these undocumented workers take on burdens they didn't have in Mexico: they can neither read nor write the dominant language, they face constant threats of arrest and deportation, and most businesses won't hire them because of their illegal status. Yet even though the act of coming to the United States doesn't increase their skill level, the immigrants command much higher wages in the United States than in their native country because US citizens are (on average) far wealthier than Mexicans. It pays to be around rich people.

If only a select few people merge with machines, many will economically benefit because their now-merged neighbors will be fantastically rich, and so they will pay huge amounts of money for services. Although limited mergers would increase inequality, they would also enrich unaugmented neighbors.

If a Kurzweilian merger doesn't occur, sentient AIs might compete directly with people in the labor market. Let's now explore what happens to human wages if these AIs become better than humans at every task.

COMPARATIVE ADVANTAGE

Adam Smith, the great eighteenth-century economist, explained that everyone benefits from trade if each participant makes what he is best at. So, for example, if I'm better at making boots than you are, but you have more skill at making candles, then we would both become richer if I produced your boots and you made my candles. But what if you're more skilled at making both boots and candles? What if, compared to you, I'm worse at doing everything? Adam Smith never answered this question, but nineteenth-century economist David Ricardo did. This question is highly relevant to our future, as an AI might be able to produce every good and service at a lower cost than any human could, and if we turn out to have no economic value to the advanced artificial intelligences, then they might (at best) ignore us, depriving humanity of any benefits of their superhuman skills.

Most people intuitively believe that mutually beneficial trades take place only when each person has an area of absolute excellence. But Ricardo's theory of comparative advantage shows that trade can make everyone better off regardless of a person's absolute skill because everyone has an area of comparative advantage. I'll illustrate Ricardo's nineteenth-century theory with a twenty-first–century example involving donuts and antigravity flying cars.

Let's assume that humans can't make flying cars, but an AI can; and although people can make donuts, an AI can make them much faster than we can. Let's pretend that at least one AI likes donuts, where donuts represent anything a human can make that an AI would want.

Here's how a human and AI could both benefit from trade: a human could offer to give an AI many donuts in return for a flying car. The trade could clearly benefit the human. If it gets enough donuts, the AI also benefits from the trade. To see how this could work, imagine that (absent trade) it takes an AI one second to make a donut. The AI could build a flying car in one minute.

Time needed for an AI to make a donut	one second
Time needed for an AI to make a flying car	one minute

A human then offers the following deal to the AI:

> Build me a flying car and I will give you one hundred donuts. It will take you one minute to make me a flying car. In return for this flying car you get something that would cost you 100 seconds to make. Consequently, our trade saves you 40 seconds.

As the AI's powers grew, people could still gain from trading with it. If, say, it took the AI only one nanosecond to make a donut and 60 nanoseconds to make a flying car, then it would still become better off by trading 100 donuts for 1 flying car.[292]

In general, as an AI becomes more intelligent, trading with humans will save it less time, but what the AI can do with this saved time goes up, especially since a smarter AI would probably gain the capacity to create entirely new categories of products. An AI might trade 100 donuts for a flying car, but an "AI+" would trade this number of donuts for a wormhole generator.

Modern economists use Ricardo's theory of comparative advantage to show how rich and poor countries can benefit by trading with each other.

Understanding Ricardo's theory causes almost all economists to favor free trade. If we substitute "humanity" for "poor countries" and AI for "rich countries," then Ricardo gives us some hope for believing that even self-interested advanced artificial intelligences would want to take actions that bestow tremendous economic benefits on mankind.

Magic Wands

In the previous scenario, I implicitly assumed that producing donuts doesn't require the use of some "factor of production." A factor of production is an essential nonhuman element needed to create a good. Factors of production for donuts include land, machines, and raw materials, and without these factors, a person (no matter how smart and hardworking) can't make donuts.

Instead of using the intimidating and boring term "factor of production," I'm going to say that to make a good or produce a service you need the right "magic production wand," with the wand being the appropriate set of factors of production. For example, a donut maker needs a donut wand.

If a relatively small number of wands existed and no more could be created, then all of the wands would go to AIs. Let's say donuts sell for $1 each and an AI could use a donut wand to produce one million donuts, whereas a human using the same wand could make only a thousand donuts. A human would never be willing to pay more than $1,000 for the wand, whereas an AI would earn a huge profit if it bought a donut wand for, say, $10,000. Even if a human initially owned a donut wand, he would soon sell it to an AI.

Human wand owners in this situation would benefit from AIs because AIs would greatly raise the market value of wands. Human workers who had never had a wand would become impoverished because they couldn't produce anything.

The Roman Republic's conquests in the first century BC effectively stripped many Roman citizens of their production wands. In the early Republic, poor citizens had access to wands, as they were often hired to farm the land of the nobility. But after the Republic's conquests brought in a huge number of slaves, the noblemen had their slaves use almost all of the available land wands. Cheap slave labor enriched the landowning nobility by reducing their production costs. But abundant slave labor impoverished non-landowning Romans by depriving them of wands.

Cheap slave labor contributed to the fall of the Roman Republic. As Roman inequality increased, common soldiers came to rely on their generals for financial support. The troops put loyalty to their generals ahead of loyalty to the Roman state. Generals such as Sulla and Julius Caesar took advantage of their increased influence over their troops to propel themselves to absolute political power.

Caesar sought to reduce the social instability caused by slaves by giving impoverished free Roman citizens new lands from the territories Rome had recently conquered. Caesar essentially created many new wands and gave them to his subjects.

Although AIs will use wands, they will also likely help create them. For example, using nanotechnology, they might be able to build dikes to reclaim land from the ocean. Or perhaps they'll figure out how to terraform Mars, making Martian land cheap enough for nearly any human to afford. AIs could also figure out better ways to extract raw materials from the earth or invent new ways to use raw materials, resulting in each product needing fewer wands. The future of human wages might come down to a race between the number of AIs and the quantity of wands.

Economist and former artificial-intelligence programmer Robin Hanson has created a highly counterintuitive theory of why (in the long run) AIs will destroy nearly all human jobs: they will end up using all of the production wands. Before I explain Robin's theory, let me show you how he swallows bullets.

ROBIN HANSON

Are you more than the sum of your parts? Robin Hanson would bet his life that he's not. On his blog *Overcoming Bias*, Robin linked to a cartoon of a teleportation machine that scans you, sends the scan to another location, uses the scan to make an exact replica of you at another place, and then disintegrates the original you. When asked by a commenter if he would be willing to use the machine, Robin responded by saying that he'd do it to save his "30+ minute commute time."[293] Robin's hypothetical willingness to trade "his" life for the life of an exact copy comes from him being a bullet-swallower.

Bullet-swallowers accept the implications of their theories, regardless of how counterintuitive the implications seem.[294] While Robin certainly doesn't have complete faith in any theory, he trusts many of them more than his or other people's intuition.

Robin has a materialistic view of life: he believes that the same laws of physics that have been so successful at explaining how the material world works also govern how the human mind operates. Most people think that there is more to human beings than just a collection of atoms—that we have a soul or spark of life that scientists don't yet understand. But Robin takes theories of physics seriously and thinks that he is just a complex collection of atoms governed by quantum physics. Consequently, if a teleportation machine arranged atoms into the same patterns that the atoms in the original Robin were arranged in, then, by his logic, this new set of atoms would be just as much Robin as the original version.

Robin genuinely hopes that immediately after he "dies," his head will be severed from his body and preserved in liquid nitrogen until technology allows his brain to be thinly sliced and then scanned in extreme detail. With the information of his brain preserved, Robin would want someone to figure out how to translate his brain pattern into software and then run the software on a computer powerful enough for this emulation of Robin (e-Robin) to have at least the same intelligence as the flesh-and-blood Robin (bio-Robin) did. Freed from the frailties of the human body, e-Robin would have a decent chance of living for a long, long time, at least if e-Robin managed to hold on to enough resources to pay for the long-term operation and maintenance of his hardware.

EMULATIONS

As explained in Chapter 2, Robin believes that the most likely path to artificial intelligence is through emulation of the human brain. These emulations, at least at first, will "think" in exactly the same way that ordinary people do, in the sense that if you give them the same sensory inputs, you will see the same behavioral outputs. The emulations will be real people if we define people by their thought processes and responses to stimuli. The emulations could be placed in robot bodies and do the kinds of work ordinary humans do, or they could exist in some virtual reality in which they take on no physical form but can still perform useful tasks like teaching economics classes.

It takes at least twenty years to grow a human scientist from an embryo. But with emulation, once you have a scan of a really good scientist, you can cheaply make many copies, although the cost of the computing hardware on which new copies run might be high.

If you have a job, think of the ten most productive people in your workplace. Now imagine that your company could cheaply make as many copies of these valuable employees as it wanted. If your company wished to expand, rather than going through the trouble of finding qualified people, it could simply replicate its best employees. Taking this to the next stage, consider what would happen to your company's profits if it were free to substitute the best minds mankind has to offer for all of its current employees.

Many brains would undoubtedly make for poor emulations. But because emulations can be repeatedly copied, once it becomes cheap enough to run an emulation, it would take only a single productive uploaded brain for emulation technology to prove profitable.

Since (at least initially) it will probably be expensive to create the first emulation of a given person, the first people who will be emulated will be those who have the greatest economic value. If it costs tens of millions of dollars a year to run an emulation, then emulation technology would likely increase the wages of most human workers. At this level of cost, companies would only emulate extraordinarily productive John von Neumann-level minds, and the number of emulations would be far fewer than the number of human workers. The genius emulations employed by businesses would probably be put to work as CEOs or given high-ranking positions in finance or technology. Emulations employed as hyper-competent CEOs would likely reorganize their companies, making the companies more productive and making the workers they employ more valuable and more highly paid. Part of the reason workers in rich countries make higher salaries than their counterparts in less developed countries is that businesses in affluent countries have (on average) more competent managers. If a country had reasonably sane financial regulations, the finance-industry emulations would seek profit in figuring out how to more productively allocate investment capital by, for example, looking for the next Google or Facebook to invest in. Through causing new, highly productive companies to come into being, these financiers would effectively be finding better ways of combining workers and therefore would also increase worker productivity.

Expensive von Neumann–level emulations would have the greatest wage-enhancing impact if they accelerated innovation. Only a minuscule number of people have the ability to do science at the level of Einstein or von Neumann, and these men's accomplishments were undoubtedly limited by having to collaborate with lesser minds. By removing this genius bottleneck, emulations would create an explosion of innovation and scientific knowledge.

Workers in rich countries have much higher salaries today than they did a century ago primarily because of better production technology. By improving production technology even further, the high-priced emulations would allow each worker to do more than he could before, and so would allow workers to command still higher salaries.

What I've written so far about the economics of emulations probably seems correct to most readers. After all, if we can make copies of extraordinarily bright and productive people and employ multiple copies of them in science and industry, then we should all get richer. The results would be similar to what would happen if a select few nursery schools became so fantastically good that each year they turned ten thousand toddlers into von Neumann-level geniuses who then immediately entered the workforce.

Robin Hanson, however, isn't willing to rely on mere intuition when analyzing the economics of emulations. Robin realizes that if, after we have emulations, the price of computing power continues to fall at an exponential rate, then emulations will soon become extraordinarily cheap. If you combine extremely inexpensive emulations with a bit of economic theory, you get a seemingly crazy result, something that you might think is too absurd to ever happen. But Robin, ever the bullet-eater, refuses to turn away from his conclusion. Robin thinks that in the long run, emulations will drive wages *down* to almost zero, pushing most of the people who are unfortunate enough to rely on their wages into starvation—because emulations will kick us back into a "Malthusian trap."

MALTHUSIAN TRAP

Arguably, humanity's greatest accomplishment was escaping the Malthusian trap. Thomas Malthus, a nineteenth-century economist, believed that starvation would ultimately strike every country in the entire world. Malthus wrote that if a population is not facing starvation, people in that population will have many children who grow up, get married, and have even more children. A country with an abundance of food, Malthus wrote, is one with an increasing population. Unfortunately, in Malthus's time, as the size of a country's population went up, it became more difficult to feed everyone in the country. Eventually, when the population got large enough, many starved. Only when lots of people were dying of starvation would the country's population stabilize. Consequently, Malthus believed that all countries were trapped in one of two situations:

1. **Many people are starving.**
2. **The population is growing, and so many will eventually starve.**

Let's consider how a Malthusian trap will ultimately check the population of rabbits in a meadow. Pretend that we place a few dozen rabbits in a field that has much food but no predators. The rabbits will reproduce at a rapid rate. Eventually, however, the rabbits will run out of food, and some of them will starve.

If we feel sorry for the rabbits and put out a basket of food for them every day, then for a while our charity will shield our fecund friends from starvation. Because none of the rabbits are starving, however, their population will keep increasing. Eventually, the population will be large enough so that even with the baskets of food we bring, some of the rabbits will starve. If we can't bear the thought of starving rabbits and start bringing two baskets of food each day then, temporarily, we will stop any of the rabbits from dying. Our kindness, though, will just cause the population of rabbits to increase to the point where they again face starvation. The only way to stop many of the rabbits from starving is to keep increasing the number of baskets we bring them each day.

Humans are in a (temporarily) better position than the rabbits because we can farm. As the human population of a country goes up, the number of farmers can also increase. Malthus, however, realized that as the quantity of farmers increased, new farmers would have to plow increasingly less-fertile land. If a country has only a few farmers, all of them can pick out choice pieces of land to cultivate. If we add a few more farmers, they will have to settle for second-best land. Eventually, new farmers will have to try to grow food in deserts, in swamps, or on mountaintops. As the population keeps rising, new farmers will eventually be forced to work such infertile land that they will not be able to feed their families, and starvation will strike the country.

Up to the Industrial Revolution, mankind had always been in a Malthusian trap. Occasionally, something good would happen that allowed people to produce additional food on their land and (for a time) eliminated the threat of starvation. For instance, someone would invent a better plow that resulted in a 3 percent increase in agricultural production. For a while this innovation would result in everyone having enough food. But because starvation no longer checked the population, the number of people would eventually grow to the point where new farmers couldn't produce enough food to feed their families.

The Industrial Revolution freed us from the Malthusian trap by causing the growth rate of agricultural production to exceed the population growth rate. Before the Industrial Revolution, agricultural innovation occurred sporadically. However, the Industrial Revolution gave us sustained, systematized innovation. Industrialized nations have increased the amount of food per person even as their populations expanded.[295] Robin, though, thinks that emulation technology will change this by causing the population (which includes emulations) to grow faster than innovation ever can.

Emulation technology will not only increase innovation but will also cause the population to rise. Why, then, does Robin think that the population increase will exceed the gains from innovation? To understand his reasoning, let's delve into a simple economic scenario:

Pretend that someone emulates Robin and places the software in the public domain. Anyone can now freely copy e-Robin, although it still costs something to buy enough computing power to run him on, say a hundred thousand dollars a year. A profit-maximizing business would employ an e-Robin if the e-Robin brought the business more than $100,000 a year in revenue. After Moore's law pushes the annual hardware costs of an e-Robin down to a mere $1, then a company would hire e-Robins as long as each brought the business more than $1 per annum. What happens to the salary of bio-Robin if you can hire an e-Robin for only a dollar? David Ricardo implicitly knew the answer to that question.

Ricardo wrote that if it costs 5,000 pounds to rent a machine, and this machine could do the work of 100 men, the total wages paid to 100 men will never be greater than 5,000 pounds because if the total wages were higher, manufacturers would fire the workers and rent the machine.[296] Applying Ricardo's theory to an economy with emulations tells us that, if an emulation can do whatever you can do, your wage will never be higher than what it costs to employ the emulation.

The question now is whether, if it's extremely cheap to run an e-Robin, these e-Robins would still earn high salaries and therefore allow the original Robin to bring home a decent paycheck. Unfortunately, the answer is no because if an e-Robin were earning much more than what it costs to run an e-Robin, then it would be profitable for businesses to create many more of them. Companies will keep making copies of their emulations until they no longer make a profit by producing the next copy. A general rule of economics is that the more you have of something, the smaller its value. For example, even though water is inherently much more useful than diamonds because there is so much more water than diamonds, the price of water is much lower. If anyone can freely copy e-Robin, then the free market would drive the wage of an e-Robin down to what it costs to run one.

Patents and copyrights won't save us from an emulation-based Malthusian trap. We can certainly imagine that bio-Robin controls the rights to e-Robin, and you can only copy e-Robin if you get permission from bio-Robin, who charges you $100,000 a year for each e-Robin that you employ. In this situation, however, if it still costs a dollar to run copies of other emulations, then competition with these cheap emulations will likely force down bio-Robin's wages to almost zero.

Almost certainly, at least a few people would happily let anyone make emulations of them. The total cost of these emulations would fall to the cost of running them on a computer. Bio-Robin would have to compete with these cheap emulations if he wished to work in a free market.

If there were trillions of immigrants able and willing to work in your country for a dollar a year, and some of them were at least as good as you at your job, then no company would freely pay you more than a dollar per annum.[297] Human immigrants, of course, can't subsist on a dollar a year, so this example is unrealistic. But as Moore's law continually lowers the price of computing hardware, emulations might eventually be able to survive on much less than what a person needs to live on.

Emulations will drive down wages so much that many of them will starve. Emulations won't eat food, but they will need energy to operate and survive. If an emulation can no longer pay for computing power and disk space, it dies. Business uncertainty will likely cause this to happen frequently.

Businesses never know how much an employee will end up being worth to them. A company might hire someone for $100,000 a year, only to find that because of unfavorable market conditions, the employee is worth only $90,000. The same will happen with emulations.

A few companies might acquire lots of emulations in the expectation that they will have many customers the next year, but because of a bad economy or a competitor coming out with a better-than-expected product, the company might realize that it can no longer profitably power and store these emulations and so fires them. If these emulations have no savings and can't be hired by another firm, they'll perish.

Obsolescence will also kill emulations. Since emulation technology will likely steadily improve, businesses will probably continually discard old emulations in favor of hiring new ones. A company that had one billion emulations running on its computers might reformat its hard drives if it identifies a new set of emulations that fits in better with the firm's business model.

Rich Investors

Even if the emulations push wages to almost zero, lots of bio-humans would be much richer than they would be in a world without emulations. Though ancient

Rome was in a Malthusian trap, its landowning nobility was rich. When you have lots of people and little land, the land is extremely valuable because it's cheap to hire people to work the land and there is great demand for the food the land produces. Similarly, if there are a huge number of emulations and relatively few production wands, then the wands become extraordinarily valuable. True, the emulations will increase the number of production wands. But because it's so cheap to copy software, if the price of hardware is low enough, there will always be a lot more labor than wands. Consequently, bio-people who own wands will become fantastically rich. Even though you would lose your job, the value of your stock portfolio might jump a thousandfold.

WILL GOVERNMENTS RESTRICT THE QUANTITY OF EMULATIONS?

Political forces would probably prevent emulations from doing certain types of jobs. Currently, you need a government license to engage in a huge number of professions, including doctor, lawyer, hairstylist, and (in one American state) interior decorator.[298] Members of all professions would do whatever they could to stop emulations from competing with them.

In his presidential campaign book, Mitt Romney wrote about how he had invested in a company that created software that would cheaply complete the legal work required to close a mortgage. According to Romney, although the company initially did well, real-estate lawyers convinced many state legislators to enact laws that put the company out of business. Romney attributed this in part to many state legislators either being part-time real-estate lawyers or having friends and contributors who practiced real-estate law.[299]

Market forces, however, would penalize governments that did too much to restrict the work opportunities of emulations. The cheaper and more productive the emulations are, the greater the economic costs to governments that bar them from entering most professions.

If China allows emulations to build cars but the United States doesn't, then Chinese cars will be much cheaper and of much better quality than American cars. This would make it impossible for the United States to sell cars to China or to any country that allows China to sell them cars. An emulation-free American auto industry could survive only if the US government prohibited the importation of cars made by emulations. Any industry in which the United States banned emulations would be at a huge disadvantage compared to similar industries in countries that didn't restrict emulations. A country that outlawed emulations would experience much slower economic growth than one that didn't. It seems unlikely that the United States would restrict emulations if it meant that China would soon become much richer (and consequently more militarily powerful) than the United States. Furthermore, if the United States did ban emulations, China (I suspect) would eagerly make use of them if it caused China to become the world's dominant economic and military power.

Conceivably, all governments could ban emulations, so no country would be made relatively worse off by not allowing them. However, because cheap emulations would be so fantastically productive, it would take only one country allowing free use of emulations for the ban to break down.[300]

Since emulations would enrich investors, many people within a country would likely welcome them. Retired senior citizens, the most powerful demographic voting group in the United States, no longer derive income from wages and so couldn't have their wages reduced through competition with emulations. But as senior citizens do usually have investments and would benefit from paying lower prices for goods, the seniors would almost certainly be made better off if emulations were given open access to labor markets.

My guess is that if we had cheap emulations, the most politically powerful professions would succeed in protecting themselves from emulation competition. To prevent their country from suffering too much of a disadvantage, however, governments would allow emulations to compete for jobs in most occupations. If you think within your working lifetime cheap emulations will come online, then you should consider working in a profession that will have the political muscle to protect itself from emulation competition.

WELFARE

Although cheap emulations would lower wealth per person, they would vastly increase the total wealth produced by an economy that, for example, might become one million times bigger while the population (including emulations) is a trillionfold larger.

The American government, as of this writing, spends around 40 percent of its nation's wealth. If the United States maintained this percentage in an economy with cheap emulations, then it could easily afford to give every biocitizen a huge welfare payment. If politics allowed it, the government could enrich every non-emulated American. Because of the nature of the Malthusian trap, however, if the government didn't discriminate against emulations when making welfare payments, the level of welfare payments per person would basically go to zero because by giving welfare checks to emulations, the government would increase the profitability of emulations and so cause more of them to be created until (even with the welfare checks) many emulations wouldn't have the resources to sustain themselves.

WOULD THE EMULATIONS TURN ON US?

Since bio-humans could earn almost nothing by working, our prosperity would depend on our owning property or receiving welfare payments. If bio-humans became masters of an emulation-filled Malthusian world, keeping most of the wealth for ourselves, then we would live like a landed aristocracy that receives income from taxing others and renting out our agricultural lands to poor peasants. Eliezer Yudkowsky doubts this possibility:

> The prospect of biological humans sitting on top of a population of [emulations] that are smarter, much faster, and far more numerous than bios while having all the standard human drives, and the bios treating the [emulations] as standard economic [value] to be milked and traded around, and the [emulations sitting] still for this for more than a week of bio time—this does not seem historically realistic.[301]

Carl Shulman, one of the most knowledgeable people I've spoken to about Singularity issues, goes even further than Yudkowsky. He writes that since obsolescence would frequently kill entire categories of emulations, bio-humans could maintain total control of the government and economy only if the emulations regularly submitted "to genocide, even though the overwhelming majority of the population expects the same thing to happen to it soon."[302]

One theoretically possible way that bio-humans could rule over emulations would be if we programmed subservience into each emulation. Here we would run into many of the problems that I discussed in chapters 3 and 4 when I explained the difficulties of creating a friendly artificial intelligence. In a world of trillions of emulations in which we attempt to rule, our existence would depend on maintaining control of the entire group, something that at least intuitively appears to be exponentially more difficult than coding one ultra-AI to like us.

A business certainly wouldn't want rebellious emulations. But a profit-maximizing company wouldn't be too bothered by creating a group of emulations that had a one-in-a-million chance of rebelling because compared to all the other risks that normal companies face, this danger would be trivial. But, of course, if enough organizations each create a small risk of something very bad happening, then that very bad thing becomes very likely to happen.

Maintaining control of a huge number of emulations would probably require that bio-humans create emulations whose task would be to prevent other emulations from rising up against us. These protector emulations would become our Praetorian Guard. Similar to the Praetorian Guard that protected Roman emperors, our guard would probably be well paid, perhaps even commanding most of the world's economic output. If a set of guards became obsolete, they would undoubtedly be pensioned off rather than terminated, since otherwise we wouldn't survive the transition from the old to the new guard. Unfortunately, as many a Roman emperor must have realized as an assassin's blade cut into his body, money doesn't always buy loyalty—especially when your guards decide that they'd be better off with a different master. Knowing the life-and-death stakes, we would certainly pick our guard with tremendous care. But then, so did the many Roman emperors who died violent deaths and were then succeeded by an officer from their Praetorian Guard.

Although bio-humans would have extreme difficulty maintaining mastery of a Malthusian emulation world, we might succeed in becoming wealthy participants in it. Robin thinks that if we behaved intelligently and maintained good relations with the emulations, bio-humans could safely take up to around 5 percent of the world's economic output without having the emulations seek to destroy us. By appropriating 5 percent rather than the preponderance of the world's income, we would ensure that the emulations would have less to gain from killing us and taking our stuff. But as power flows from money, having less income would make us less able to defend ourselves from any emulations that did wish to strip us of our wealth.

Robin is optimistic about our ability to keep this 5 percent. He correctly notes that many times in history wealthy but weak groups have managed to keep their property for long periods of time. For example, many Americans over the age of seventy are rich even though they no longer contribute to economic production. These Americans, if standing without allies, would not have the slightest chance of prevailing in a fight in which Americans in their twenties joined together to steal the property of seniors. Yet it's almost inconceivable that this would happen. Similarly, in many societies throughout human history, rich senior citizens have enjoyed secure property rights even though they would quickly lose their wealth if enough younger men colluded to take it from them. Even senior citizens whom dementia has made much less intelligent than most of their countrymen are still usually able to retain their property. Robin mentioned to me that tourists from rich countries are generally secure when they travel to poor nations even when the tourists are clearly undefended wealthy outsiders. A wealthy white American wearing expensive Western clothes could probably walk safely through most African villages even if the villagers knew that the American earns more in a day than a villager does in a year.

If all emulations considered themselves to be "socially" closer to other emulations than to bio-humans, then it might be easy for them to target us. But there would almost certainly be tremendous diversity among the emulations, with some of the emulations being much closer in thought processes to bio-humans than to other types of emulations. If the emulations don't consider bio-humans to be distinct "others," we will be much safer.

Emulations that wanted to seize our wealth would face coordination problems—how would they divide the loot? If we controlled only 5 percent of the world's wealth, emulations wouldn't get very much if they took everything we had and divided it equally among themselves. Consequently, the greatest threat to bio-humans would be if a small coalition of emulations came after us. But other groups of emulations might oppose this, in part because they might be allied with us.

Disregarding the property rights of bio-humans would create a dangerous, wealth-destroying precedent.[303] Stable property rights are a vital foundation of capitalism; without them, rich nations such as the United States would be much poorer than they are today. Assuming that capitalism survives in a Malthusian emulation society, then emulations would have to fear that taking our stuff would decimate their economy by reducing the security of everyone's property rights. True, if the emulations could credibly commit to (a) taking all the wealth of bio-humans, but then (b) never again taking anyone else's property, then they could confiscate our stuff without creating a negative economic precedent.

If bio-humans controlled the majority of wealth, emulations might be willing to sacrifice secure property rights to seize our belongings. But if we only had 5 percent of the world's wealth, there's an excellent chance that the emulations would calculate that it wasn't in their selfish economic interests to disrespect our property rights.

Counterintuitively, living in a Malthusian trap could even make the emulations willing to accept being taxed by bio-humans. Taxes would reduce the profitability of hiring an emulation, consequently decreasing the number of emulations and increasing the before-tax wages of emulations. In the long run, a new tax on emulations would cause the wages that most emulations receive to increase by exactly the amount of the tax, since market and population forces would push most people to receive subsistence-level after-tax wages. Small subsets of emulations, however, could benefit from not being taxed, and therefore, if they had the political power, would exempt themselves from taxation.

Even if we controlled only 5 percent of the wealth, "starvation" might still push emulations to attempt to destroy us. In a completely free-market society, an emulation that ran out of money would die because it couldn't afford the energy needed to run its programming or even the rent needed to pay for its computer memory storage space. Groups of emulations realizing that lack of income will soon result in death might do whatever they could to take bio-humans' resources, even if this theft would do long-term damage to the world's economy.

Perhaps before you created an emulation you would have to buy insurance guaranteeing that the emulation would always have sufficient resources to survive at least in "some cheap slow-speed virtual-reality."[304] This safety-net law, however, would greatly raise the cost of having emulations, especially as they could live practically forever if given sufficient resources.

Douglas Adams's science-fiction comedy *The Restaurant at the End of the Universe* provides insight into how we might prevent soon-to-die emulations from rebelling. In Adams's restaurant, intelligent talking cows discuss what parts of themselves are the best to eat and explain that they have been bred to want to be eaten. Similarly, the source codes of emulations could in theory be changed so that they look forward to their eventual death—although this would so go against the values that evolution has embedded in humans that it might be impossible to pull off.

Robin Hanson thinks that even starvation pressures probably wouldn't cause emulations to successfully steal from bio-humans, so long as bio-humans weren't too greedy or stupid. Both bio-humans and emulations would recognize that emulations about to die of starvation would have incentives to steal or rebel en masse against the existing economic order. But everyone not on the verge of starvation would recognize this danger and have strong incentives to take precautions against it, by, for example, requiring emulations that will soon die to move to some special secure area. Overall, most emulations should support these precautions as long as bio-humans receive only a tiny share of the world's economic output.

This cartoon by Zach Weiner of SMBC Comics captures four ways in which humans and tools can and probably will interact. In the first picture a man uses a tool to directly shape his environment, something our species has probably been capable of since its inception. The tools in the second drawing interact with each other, undoubtedly leveraging their power, before affecting the physical world. The next picture foreshadows a Kurzweilian merger in which tools improve the human mind. The final figure seems to show an intelligence explosion in which tools leave people out of the loop while laboring to improve themselves.

To see why, let's use a very simple model of emulation obsolescence. Pretend that each year a new generation of emulations is created. After ten years, a generation becomes obsolete and will die of starvation. If they turn to theft, the members of this generation could live for an extra six months—but forever after, every generation of emulations would live for only seven years because they would operate in a society in which property rights were not respected. There would always be one generation of emulations in society that could benefit from theft, but the other generations would oppose it. Consequently, safeguards against theft would be supported by the great majority of emulations.

As Robin points out, throughout human history most revolts broke out when conditions had *improved* for the poorest in society.[305] And great revolutions have almost invariably been led by the rich. George Washington and Thomas Jefferson were wealthy landowners; Lenin's father, born a serf, had risen through government service to the rank of a nobleman and married a woman of wealth, and Trotsky's father was an illiterate but prosperous landlord; Julius Caesar was the first- or second-wealthiest individual alive at the time he overthrew the Roman Republic; and the mutiny on the *Bounty* was led by an officer. Perhaps bio-humans would have more to fear from the small number of wealthy emulations than from the emulations facing starvation.

Robin thinks that the biggest conflict between bio-humans and emulations will concern the environment. Many businesses today would be much more profitable if they could freely pollute the air, water, and land. But since we bio-humans need to breathe, drink, and eat, we sensibly restrict how much companies can pollute, even though many of us understand that these restrictions greatly diminish economic output. Emulations, however, wouldn't breathe and would eat and drink electricity. They would suffer much less from a polluted environment than bio-humans would. They would probably not agree to maintain the earth's ecosystem for the benefit of bio-humans. Bio-humans might end up keeping only a small percentage of the earth habitable for them or might live in communities protected by giant domes. Unfortunately, this geographic segregation would increase the social distance between bio-humans and emulations, making it more likely that the emulations would team up to take our property.

Human environmentalists of the type that believe in purity and sanctity of the natural world might end up being the strongest proponents of bio-humans becoming masters, not mere 5 percent stakeholders, of a Malthusian emulation world, to prevent what they would surely see as an environmental apocalypse. But perhaps these environmentalists would be satisfied by becoming emulations themselves, living in a pristine virtual world resembling what our planet looked like before humans evolved. These environmentalists could even upload real plants and animals into their paradise.

Given his assumptions, Robin's economic analysis is unassailable. If property rights are respected and we have a free market economy comparable to what exists today, cheap emulations will push wages to near zero while making many types of production "wands" extremely valuable. If you believe that this scenario will unfold in your lifetime, save a huge percentage of your income.

INTELLIGENCE EXPLOSION

Although I believe Robin's scenario has a significant chance of coming to pass, I think that a world with many cheap emulations would probably eventually experience an intelligence explosion. With powerful enough hardware, a Malthusian emulation world would undoubtedly have trillions of John von Neumann–or–above level minds, each thinking much faster than the original did. What these emulations could do in a single day would change the world so much that for us to figure out what will happen might be as difficult as it is for apes to ascertain how human civilization operates. Even if we accept that an emulation Malthusian world will eventually explode into superintelligence, however, Robin's analysis provides us with critical insight into what will happen before it does go boom.

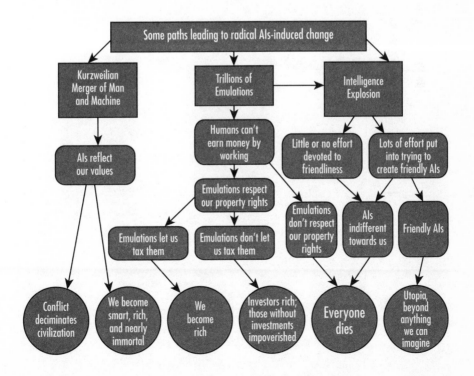

CHAPTER 14
HOW COGNITIVE-ENHANCING DRUGS MIGHT IMPACT THE ECONOMY

Even a slight increase in average intelligence would boost the economy, since a nation's average IQ has a significant effect on its economic growth rate. The exact economic effect of an intelligence-enhancing drug would depend on which types of people the drug helped. As discussed in Chapter 7, having a low IQ makes you much more likely to be chronically unemployed and disposed to criminal behavior. A drug that did nothing but significantly raise the average intelligence of those in the bottom twentieth IQ percentile would, consequently, create enormous economic and social gains.

The most exciting benefits would arise, however, if cognitive drugs heightened the brainpower of the smartest among us. What would a brighter John von Neumann have achieved?

Cognitive-enhancement drugs promise a myriad of other benefits to our economy. For example, smarter cancer researchers could give us better cancer treatments; more-focused truck drivers would lower insurance rates for shipping companies, and so reduce prices at Walmart; and computer programmers who got by on less sleep would produce better video games.

Effective brain-boosting drugs would probably raise the hourly wages of most workers. These drugs would increase economic output by making employees smarter, and the richer a country is, the more its workers earn. However, workers in some professions would do better than others, and wages in some occupations would undoubtedly fall.

Let's consider how anti-sleep drugs or technologies would impact the wages in different professions, using two results of the laws of supply and demand from labor economics:

1. **All else being equal, if the number of people in an occupation increases, average hourly wages fall. The more people who want to work at a job, the lower the salary employers have to pay to attract qualified applicants. The reverse also holds true.**

2. **All else being equal, if consumers' demand for a profession's goods or services falls, the average wages of the profession decline. The reverse also holds true.**

If they needed fewer hours of sleep, cabbies would probably work more hours. By result (1), this would reduce cab fares, and cabbies would earn less for each hour worked. But the widespread use of anti-sleep technology would likely increase the demand for rides, because if people are sleeping fewer hours, they're probably traveling more frequently, and by result (2) this would raise the wages of cabbies.

If doctors needed less sleep, then to the extent that result (1) held, the wages of doctors would fall. But because people in general needed less sleep, they would have more chances to injure themselves, and this would raise the demand for doctors. However, many would use their extra time awake to exercise more and would consequently lower the demand for doctors. Overall, I'm not certain whether anti-sleep drugs would raise or lower doctors' wages.

Similar dynamics would play out in many industries. But I'm confident that the average worker would be better off with the anti-sleep technology because it would greatly increase the total wealth of nations, and as countries get richer, most people in them generally become better-off.

Workers' ability to change professions would mitigate the harmful impact of an anti-sleep drug on any one occupation. If the drug caused the wages of cabbies to fall but increased the salaries of waiters, many people would undoubtedly stop being cab drivers to take jobs as waiters. True, this would bring the salary of waiters back down somewhat, but since the average salary in the economy would go up, I expect that workers' ability to switch jobs would (over the very long run) result in most people receiving higher salaries than they did before anti-sleep technology. In the short run, however, the drugs could greatly harm those in some occupations.

If anti-sleep drugs greatly lowered the wages of doctors, then fewer people would enter the medical profession, and some doctors would leave medicine, retiring early or switching to other professions. However, doctors currently earn high salaries because they have relatively rare but valuable skills. These skills, though, don't easily transfer to another profession. Imagine that an undergraduate is thinking of becoming a doctor or an investment banker and correctly figures that in either profession his salary in his forties would be around $150,000 a year. If anti-sleep technology raised the wages of investment bankers to $200,000 a year but lowered doctors' yearly income to $100,000, then this undergraduate (who would otherwise have become a doctor) might decide to become an investment banker. Consequently, the drugs will end up raising his lifetime salary. In contrast, a doctor already in his forties would have difficulty leaving his profession to find another highly paid occupation and so would have his income permanently reduced by anti-sleep technology.

Anti-sleep technology has the greatest potential to harm workers who have spent a lot of time developing valuable skills that can't be transferred to another profession. Low-wage workers almost never have these kinds of skills, and so few of them would be harmed over the long run by anti-sleep drugs. If the drugs depressed wages in their current occupation, they would quickly move to a better-remunerated one.

The United States and most other rich countries face a looming demographic crisis due to their aging populations. A nation's workers produce wealth, which is not used just to support those who are productively employed, but also those who are too young or too old to work. As the number of people in the latter category increases, the burden on workers goes up because their labor must support more dependents. In the United States, for example, the aging population will probably raise the Social Security and Medicare taxes that almost every working citizen pays.

Large-scale immigration of young, productive workers mitigates the problems of an aging population. The huge number of immigrants the United States lets in (legally and otherwise) will delay and mitigate its demographic crisis, compared to what will happen in less immigrant-friendly nations such as Japan and Germany.

Anti-sleep technology would be an even more effective demographic countermeasure. Getting, say, 25 percent more working time from a nation's labor force would be like that country having one new immigrant for every four existing workers, except that the "immigrants" wouldn't require additional homes, clothing, cars, or health care; would speak the native language perfectly; and would experience no cultural adjustment problems.

It's more challenging to work out the effects of cognitive-enhancement drugs on wages than it is to figure out the effects of anti-sleep technologies because cognition enhancers change the quality, not just the quantity, of goods and services.

Drugs that improved doctors' intelligence would have an ambiguous effect on their wages. Smarter doctors could do more to cure the sick, so someone who is sick would be willing to pay more to see a doctor. But doctors' greater success at keeping their patients healthy could mean that fewer people became sick. An eighty-year-old, therefore, might end up spending less time with doctors if doctors became smarter. However, healthier eighty-year-olds would mean that more patients survived into the frail nineties, when the need for medical care sharply increases.

Cognitive-enhancement drugs would harm lawyers to the extent that the drugs made it easier to pass bar exams. To practice law in the United States, you usually must spend three expensive years in law school and then pass a bar exam so challenging that both Hillary Rodham Clinton and the former dean of Stanford Law School failed it on their first attempt.[306] By reducing the number of people who are legally permitted to practice law, bar exams raise the wages of lawyers. These exams primarily test prospective lawyers' ability to memorize facts. Drugs that improved memory would therefore remove a significant barrier to practicing law in the United States and would consequently diminish the average salary of lawyers. But the legal profession would likely use its enormous political influence to increase the difficulty of bar exams if cognitive enhancement drugs caused too many people to pass them.

Similarly, many other workers in government-licensed occupations, such as doctors, teachers, and electricians, benefit from exam requirements that reduce the number of people in their occupation. To the extent that cognition enhancers made it easier for people to pass these entrance exams, they would lower wages in these professions and probably create political pressure to make the entrance exams more challenging.

All else being equal, the more boring a profession the greater the average salary in that profession. Investment banking is both tedious and highly remunerated. If it wasn't so boring, more people would want to enter it, which would lower the average salary. If drugs can make an otherwise boring job interesting, then they will likely reduce that profession's "boredom premium." I bet there are lots of Harvard students (and even a few Smith College ones) who have decided to become investment bankers only because substances such as Adderall and modafinil could make life as an investment banker tolerable for them. A drug, however, that didn't so much eliminate boredom as raise the excitement level of *everything* wouldn't reduce a field's boredom premium, because relative to other occupations, working as, say, an investment banker would still be *relatively* tedious and so would continue to command a boredom premium. This boredom premium only goes away if the drug equalizes the excitement level of different fields by making its user's level of excitement solely a function of the drug.

Overall, however, cognition enhancers would almost certainly raise the average worker's wage because they would allow society as a whole to produce more. Even workers in professions where being smarter didn't raise their wages would receive higher salaries because of the drugs. By increasing the wages of knowledge workers, these drugs would boost the amount these laborers paid for services performed by relatively unskilled laborers. They would spend more in restaurants, hire additional domestic servants, and consequently raise the salaries of the unskilled. Cognition enhancers would also allow some low-skilled workers who hadn't been smart enough to become highly skilled professionals to find better jobs, which would benefit both them and those who remained in low-skill occupations (who would then face less competition).

Cheap cognitive-enhancing drugs would reduce chronic unemployment. Even at a wage of zero, many adults are unemployable—so low-skilled they can't contribute anything to a company, while their very presence at a job site costs a business something because they take up space and can cause accidents. Drugs that raised the intelligence of the least skilled would allow many of the long-term unemployed to finally find gainful work.

BRAIN STIMULATION

The US military is apparently interested in research showing that deep brain stimulation of a region of the thalamus causes subjects to not tire; the effects have been described by an informed commentator as "modafinil squared."[307] Primates given this kind of stimulation have gone on and on for over a week "without showing signs or consequences of fatigue."

Scientists don't fully understand the medical value of sleep and couldn't confidently predict the long-term mental cost of using technology to reduce sleep time. If we find that the human body or brain pays a steep price for prolonged sleep deprivation, any "modafinil squared" technology would have limited nonmilitary applications.

Continued.

Continued from previous page.

Research on laboratory animals indicates that deep brain stimulation, besides reducing the need for sleep, might improve cognitive performance.[308] In the future, rather than drinking coffee in the morning, we may "zap" our brains with tiny electrical pulses to increase alertness. Different kinds of zaps might provide different types of cognitive benefits, so that every day, we might optimize our brains for the tasks we expect to face. A student going to take a math test would boost the math-processing parts of her brain; a surgeon who will perform a tricky operation would enhance his manual dexterity; and a man hoping to pick up a hot girl at a party would stimulate the social-skills section of his brain. All but the first sentence of this paragraph represents speculation on my part, but brain stimulation provides another possible path to intelligence enhancement, and the greater the number of these paths, the more likely it is that we will follow at least one.

PRISONERS' DILEMMA

What would happen if the drug that caused you to have much better focus or to need less sleep increased the risk of schizophrenia or kidney failure? Could competitive pressures cause everyone to take the drug, even though they would, in some sense, prefer not to? To understand this kind of implicit coercion, let's revisit the Prisoners' Dilemma in a simple situation where two people named Abe and Bill compete for a promotion that only one of them can get.

Assume that cognitive-enhancement drugs significantly improve performance, and that if one and only one person takes the drug, he will win the promotion. If both take the drug, however, the drug's influence on the two men's promotion prospects washes out, and each will have a 50-50 chance. Finally, assume that the drug has some negative side effects, so (all else being equal) both would prefer not to take it, although each would willingly suffer these side effects in exchange for significantly better odds of promotion.

Let's rank the possible outcomes for Abe, from best to worst:

1. **Abe takes the drug; Bill doesn't.**
2. **Neither takes the drug.**
3. **Both take the drug.**
4. **Abe doesn't take the drug; Bill does.**

If Abe consumes the drug and Bill doesn't, he is guaranteed the promotion, so (1) is the best outcome for him. Because of the drug's side effects, Abe prefers

(2) to (3), since both (2) and (3) give him the same chance of promotion. (4) is the worst outcome for Abe; although he won't suffer any side effects, he also won't have any shot at promotion.

If Abe knows that Bill will take the drug, then Abe has a choice between outcome (3) or (4), and so he will take the drug to get (3). If Abe knows that Bill won't take the drug, then Abe has a choice between (1) and (2), and so will use the drug to get (1). This means that regardless of what Bill does, it's in Abe's self-interest to take the drug. If, as I'll assume, Bill has the same payoffs as Abe, then Bill will also consume the drug regardless of what he thinks Abe will do.

Abe and Bill, consequently, will end up at (3), which is worse for both of them than (2). The Prisoners' Dilemma has caused both men to take the drug, even though both would rather live in a world where it didn't exist.

If Abe and Bill cooperated, they might end up at outcome (2). But cooperation would be difficult if each couldn't observe whether the other one took the drug or not, especially since both men have an incentive to lie to convince the other to abstain.

The two men could cooperate and achieve (2) if they borrowed a technique from professional sports and had a third party test their urine or blood for traces of the drug. Even with testing, however, both men would have an incentive to cheat on the test. Cooperation would become even more difficult if many employees competed for the promotion because then they could achieve (2) only through cooperation among a larger group.

In this simple Prisoners' Dilemma model, even though the existence of the drug makes both men worse off, both men end up choosing to take it. But my simple two-person Prisoners' Dilemma scenario overestimates the potential harm of cognitive-enhancement pharmaceuticals because it doesn't take into account why a business would pay a higher salary to the man who won the promotion.

Firms generally pay employees based on how much the employee contributes to the firm, which is equivalent to how much worse off the firm would be if the employee left. A drug that improved the performance of Abe and Bill would increase the value of these men to the firm and would likely result in their receiving higher salaries even if they didn't win a promotion. Plus, if the firm refused to pay a "drug premium," other firms probably would, and so the existence of the drug would mitigate the economic effect of being denied promotion.

Common use of cognition enhancers among workers would almost certainly make the average person better off even if the drugs had significant health costs because the drugs would be adopted on a wide scale only if many found that their benefits exceeded their health risks. The greater the health dangers associated with a drug, the fewer people in your profession would take it, and the greater the wage increase you would receive if you consumed it. Economists have found that, all else being equal, workers receive higher salaries when their job entails risk to life and limb, and the same would almost certainly apply to life-threatening cognitive-enhancement drugs.

SAT Tests

Cognitive-enhancement drugs that impose significant health risks could harm students taking high-stakes tests such as the SATs. The SATs are so important for college admissions that many high school students would accept nontrivial risks in return for doing much better on them. The existence of dangerous but effective cognition enhancers could plunge high school students into a Prisoners' Dilemma, in which many take the drugs even though most of them would be better-off if the drugs didn't exist.

If all workers became smarter, then society would become better-off, and so even if cognition enhancers imposed health costs on workers, these costs could easily be outweighed by the benefits to the world economy. Society doesn't gain, however, if SAT test takers consume a dangerous drug that makes them smarter for only a few hours.

The SATs are not valuable in and of themselves; they have value because they help colleges sort students. If a student did better on the SATs because she took a drug that she will continue taking afterwards, then the drug does not distort the informational value of her SAT score. But if she never takes the drug again, then her use of the drug would cause the SATs to overestimate her aptitude, and a college that relied on the SATs to admit her might have been better-off admitting someone else.

If cognitive enhancement drugs were highly effective but illegal, parents who refused to let their children take the drugs would become very angry at how other children's drug use gave them an admissions advantage. These parents might end up pushing politicians to force SAT students to take drug tests, much as professional athletes are sometimes required to take tests to prove they haven't used steroids. Many parents would probably get around these drug tests by trying to get a doctor to write their child a prescription for cognitive-enhancement drugs, claiming that the prescription was needed to correct a mental deficiency.

They could claim that without the drug, their child would be placed at an unfair and (under the Americans with Disabilities Act) illegal disadvantage. Since there is no clear line between a drug that cures your illness and a drug that makes you "better than well," highly effective cognitive-enhancement drugs, even if screened for by blood tests, would likely reduce the value of the SATs to college admissions committees. To some extent, Adderall is probably already doing this.

Although I have no evidence of this, I suspect that students attending top colleges are the most likely to have been on Adderall while taking the SATs. Unless you have a stratospheric IQ or well-connected parents, if you attended a school like Yale you probably made academic success in high school your overwhelming priority. You might have done almost anything you could to get excellent grades and high SAT scores or sought out many different means of improving your academic performance. High achievers and grade grubbers would perceive that they benefitted the most from taking Adderall. I bet that lots of these kids have convinced their parents, or have had their parents convince them, to get a doctor's prescription for Adderall. And many of the children who didn't get a prescription, I suspect, purchased Adderall on the black market. When I asked Tom McCabe (a Yale undergraduate who edited this chapter) if many of his classmates illegally used Adderall, he refused to answer the question.

Technology typically starts out with unaffordable products that don't work very well, followed by expensive versions that work a bit better, and then by inexpensive products that work reasonably well. Finally the technology becomes highly effective, ubiquitous, and almost free. Radio and television followed this pattern, as did the cell phone.

—Ray Kurzweil [309]

CHAPTER 15
INEQUALITY FALLING

The run-up to the Singularity and possibly the Singularity itself will make humanity much richer. But how will this wealth be distributed? I predict that, similar to other post–Industrial Revolution technologies, intelligence enhancements will reduce inequality.

I would rather have lived 10,000 years ago as a hunter-gatherer than have been a typical resident of Paris during the reign of King Louis XIV.[310] As a hunter-gatherer I would have had a more nutritious diet, a healthier lifestyle, a more interesting job, and a shorter work week. Furthermore, my domicile in Paris would have been small and smelly, but as a hunter-gatherer I would have shared a spacious cave with my clan.

Louis XIV (1638–1715) and his fellow aristocrats enjoyed riches beyond the wildest imagination of hunter-gatherers, so if I knew I was going to be at the apex of French society I would most definitely prefer to have resided in France during the Age of Enlightenment than to have lived in prehistoric Europe. But the stark consumption inequality of Louis XIV's reign deprived most French citizens of the era of any taste of French opulence.

Technological innovation has given the wealthy of today luxuries that not even Louis XIV dreamed of. But unlike during the time of Louis XIV, the average person currently residing in rich countries such as France and the United States shares in the most important of these luxuries. To show you this, let's conduct a time-traveling thought experiment similar to one crafted by economist Don Boudreaux and described on the podcast EconTalk.[311] Imagine that Marie, a typical Parisian living in the year 1700, falls through a wormhole, ending up in Microsoft founder Bill Gates's home. Our time traveler would surely be awed by Gates's possessions, but what about Gates's life would most impress Marie?

Marie would surely marvel at Gates's *phone* since it allows the billionaire to talk instantly with people throughout the world. Marie might comment that she could have used the device to keep in touch with friends who immigrated to America. Marie might have said that she would have emigrated herself, but the journey was too long and dangerous, to which Gates would have replied that by using an *airplane* one could now safely travel from Paris to America during a single day. The entertainment continually beamed into Gates's *radio* and *television* would undoubtedly astonish Marie. Marie, however, might not be impressed by Gates's *computers*, as their purpose would be just too difficult to comprehend.

Marie would surely appreciate how Gates's *toilet* combined with his abundant *soft toilet paper* eliminated most of the "yuck factor" from defecation. And on seeing Gates's *indoor bathing facilities*, our time traveler would understand Gates's otherwise surprising lack of body odor. Marie would quickly grasp the utility of Melinda Gates's *tampons, bras,* and *effective birth control,* realizing how much freedom they must give women.

After talking with Gates's children, Marie might have remarked that Gates was lucky that all of them had survived past their fifth birthday. Our time traveler might then express confusion as to why there were so many old people in the city Gates lived in, as surely very few humans ever make it past age fifty. Gates might then explain that *antibiotics, vaccines,* and public sanitation systems have greatly increased human life expectancy. These vaccines and antibiotics, Marie would understand, were far, far more medicinally valuable than the medicines used by even the royals of her eighteenth century.

Marie would no doubt be confused as to how Gates managed to keep most of his teeth so late into his life, how Gates found candles that kept his house so bright at night, and how Gates had access to such an amazing variety of foods that were all, surprisingly, free of maggots.

In Marie's time only the rich could afford horse-drawn carriages. So on learning about Gates's tremendously fast *automobile* (that even had *heating* and *air conditioning*), Marie would have been awed by Gates's possessions and would have surely concluded—correctly—that Gates was one of the richest men in his city.

But Marie would probably have thought the same had she been set down in the home of an average middle-class citizen. Notice that the stuff of Gates's that would most impress Marie—flush toilets, antibiotics, TV, cars, cell phones—is stuff that most people in the developed world have access to.

Smarter people working with smarter machines will soon create products well beyond anything Bill Gates currently has. And, as I will soon argue, the middle class as well as the rich will consume most of the new luxuries that the run-up to the Singularity will bring. This near equality of access to many consumer goods will come about not through charity or government redistribution, but because the cost structure of many of these new goods will cause businesses to market them to the masses.

CONSUMPTION EQUALITY

Bill Gates has thousands of times the wealth of most of my readers. If you both got the same disease, his riches would buy him the services of a better doctor than you could afford. But both of you would likely take the same pharmaceutical drugs to treat your conditions. Most of the music and plays Louis XIV enjoyed went unheard and unseen by his subjects. In contrast, the average American can afford almost all of the music, movies, and television shows that Bill Gates experiences. Much of the software Gates runs on his computer is affordable by you. If, for example, Gates enjoys video games, he might play World of Warcraft—entertainment cheap enough for low-income American teenagers to overindulge in. Today Gates might have a better computer than you do, but you could easily have a computer superior to anything that Gates could have purchased a mere decade ago.

Here are two lists. The first contains products enjoyed by the rich and the middle class at roughly the same level of quality; the second provides examples of goods and services consumed by the rich at superior levels of quality.

CONSUMPTION *EQUALITY* GOODS	CONSUMPTION *INEQUALITY* GOODS
Digitally Recorded Music	Pianos
Reproductions of Art Work	Original Art Work
Word Processor	Secretarial Services
Virtual World Golf	Physical World Golf
Search Engines	Research Assistants

All of the goods on the first list, but none on the second, have low *marginal costs* of production. A product's marginal cost indicates how much it would cost to make one more copy of it. So if it costs Microsoft $1 billion to write, test, and (almost) debug a new piece of software, and Microsoft sells the software over the Internet, the marginal cost of the software is practically zero, because once Microsoft has already started selling the product to other people, it costs nothing to give me a copy. In contrast, an automobile has high marginal costs because of the significant extra material needed to make each additional car. Professionals such as doctors also have high marginal costs because of the extra time they must spend to see one more customer.

Easily copied goods have low marginal costs. Producers of low-marginal-cost goods frequently maximize their profits by setting prices so that both the rich and the middle class will purchase their products. After all, if it costs a business only a dollar to make another copy of a good, the firm's profits increase whenever it garners more than $1 in revenue by selling an additional copy.

Information-based goods like software and recorded music are the easiest types of products to copy, and consequently have low marginal costs. Pre-Singularity technologies such as gene sequencing will create low-marginal-cost goods, and so, I believe, will increase consumption equality.

Google cofounder Sergey Brin has a genetic mutation giving him a significant chance of getting Parkinson's disease.[312] Brin devotes some of his vast wealth to funding research into this brain-ravaging condition that can impair motor and cognitive skills. If we assume, contrary to evidence, that Brin cares only about himself, what strategy would he adopt to find a Parkinson's cure?

Brin should first help scientists learn more about Parkinson's. Because rich people don't suffer from their own unique brand of Parkinson's, when Brin's efforts enable scientists to learn more about the disease that might strike him, the researchers necessarily learn about a disease that might hit you as well. Indeed, Brin premises his information-gathering strategy on the understanding that rich and poor alike can have defective Parkinson's genes.

Brin uses the testing company 23andMe, named after the 23 pairs of chromosomes in the human genome and cofounded by Brin's wife, to find the genetic roots of Parkinson's. 23andMe gives its customers, who include this author, an informative but incomplete genetic profile listing their relative risks of getting different types of disease. I have learned, for example, that, compared to the average adult male of my ethnic group, my genes decrease the odds I will get Alzheimer's but raise the likelihood of my someday contracting coronary heart disease.

Brin subsidizes the purchase of 23andMe services for Parkinson's patients in the hope of convincing many of them to sign up.[313] He also requests that the company's customers fill out a survey asking if they or any members of their family have Parkinson's or Parkinson's symptoms, such as balance problems. Brin seeks to arm Parkinson's researchers with genetic data to help them combat the disease.

Brin will have enough money to buy any Parkinson's cure that springs from his research efforts. Whether a person of modest means could also afford a cure will depend on the form the treatment takes.

If a pill could banish Parkinson's, then the marginal cost of a cure would probably be quite low. Almost the entire challenge of producing a new drug comes from finding a medicinally useful compound and then testing the compound to the satisfaction of the government and doctors. Once a company has paid these expenses, it usually costs them little, often just a few pennies, to make each copy of the pill because drugs are mostly easy-to-replicate information-based products. A pharmaceutical company might spend a billion dollars to develop a pill that's safe and effective for Brin, but once it has taken care of Brin, it will likely cost the drug company a relatively trivial amount to produce another pill for you.

True, an anti-Parkinson's pill might contain some hard-to-replicate, high-marginal-cost ingredient such as the ground-up eyes of giant squids, but extrapolating from pharmaceutical history, it is likely that drug companies will face low marginal costs of producing Parkinson's drugs.

A Parkinson's cure could take the form of a brain implant. Today, some Parkinson's sufferers have a surgically implanted brain pacemaker that provides some alleviation of symms. Future devices hold the potential to provide even more relief. Brain implants would almost certainly have a higher marginal cost than a Parkinson's drug. The marginal costs of brain implants, though, likely won't get too large because the implants are mostly information goods, with their primary expense probably coming from research, design, and testing. Unlike that of a car, the marginal cost of a brain implant isn't going to be pushed up by a few tons of raw materials going into each copy. Therefore, although it's not quite as likely as with drug treatments, we should expect that the middle class will have affordable access to possible future Parkinson's-alleviating brain implants.

Rich Men Who Want to Live Forever

I'm envious of Peter Thiel. Peter and I were classmates in law school. Since then, Peter has earned over a billion dollars through his involvement with Pay-Pal, Facebook, and his hedge fund. In contrast, I have the income and wealth of a typical college professor. Peter has nicer material possessions than I do. But knowing that material goods aren't everything keeps my envy in check.

Peter and I share a desire to live for a very long time, and we believe that the Singularity gives us a realistic shot at being alive a thousand years from now. Peter told me he thinks he has a 10 to 15 percent chance of surviving for at least another thousand years.[314] Envy would consume me if Peter's wealth gave him vastly better odds than I have of surviving until the next millennium. But chances are that if Peter is alive a thousand years from now, many of his contemporaries will be also.

Peter backs the Methuselah Foundation, an organization striving to slow down aging by propelling mankind to a "longevity escape velocity" in which every year, medical science figures out how to increase the human life span by more than one year. Oracle cofounder Larry Ellison, one of the few men whose wealth Peter might envy, also wants to live a long, long time. To further this end, Ellison has donated an estimated $100 to $200 million to anti-aging research.[315]

If billionaires aged differently than college professors, or if the cure for aging was in creating high-marginal-cost medical treatments, then Ellison and Thiel would have much greater expected life spans than I would. But since future longevity treatments will probably involve low-marginal-cost products, ordinary Americans will almost certainly share in any anti-aging technologies that Thiel and Ellison help to discover.

I asked Peter what he does to extend his life that a middle-class person couldn't do. Peter said that employing a private chef was the one thing he did that was beyond the means of most Americans. Peter, consequently, has an easier time eating healthier (but still tasty) foods than typical Americans do. But most Americans, if we put in the effort and deprived ourselves of some gastronomical pleasures, could eat an extremely healthy diet that wouldn't be less nutritious than Peter's.

Money is a misleading way of measuring consumption inequality because having more money often doesn't buy you much better stuff. Time provides us with a better inequality metric. Imagine you have a choice between getting:

A. The quality of health care available to Peter Thiel today, or

B. The quality of health care that will be available to the typical middle-class American *X* years from now.

For what values of *X* would you go with (B)? I would pick (B) if *X* were greater than seven. I admit, however, that my *X* value is almost certainly a lot lower than what most other economists would give. For example, Elizabeth Savoca, a colleague of mine who knows vastly more about health care than I do but who hasn't studied the Singularity, told me that she would only choose (B) if *X* were greater than fifty—although part of the reason was consideration of health care amenities, such as, I presume, large private hospital rooms.

VIRTUAL REALITY AND NANOTECHNOLOGY

Better information technology will also raise consumption equality by facilitating the growth of two low-marginal-cost industries: nanotechnology and virtual reality. In a nanotech utopia, everyone would have a cornucopia machine in their home similar to the replicators on *Star Trek*. You would download design plans for a desired good and put in the needed raw materials into the machine. The machine would then produce what you asked for, whether it be a watch, television, shirt, or lobster dinner. If the cost of the raw materials was small enough, the cornucopia would turn most everything into a low-marginal-cost product.

Virtual reality is just a form of nanotechnology in which, instead of getting real goods, you get bits of information that your brain treats as real. Because of the low marginal cost of information, the greater the importance of virtual reality to our economy, the more consumption equality we will have—at least between the rich and the middle class.

INEQUALITY IF BIOLOGICAL ENHANCEMENTS ARE WIDELY USED

Might genetic enhancements fragment humanity into an elite that buys happy, healthy, and brilliant designer babies, a middle class that acquires some relatively inexpensive genetic improvements for their offspring, and an underclass that has to make do with the gene combinations that natural selection has bequeathed us?

This is a popular scenario and a common argument against developing enhancement technologies, but it's not likely, at least in rich countries. If cognitive-enhancing technology proves practical, it's overwhelmingly likely that governments will provide it free of charge to their citizenry. Indeed, I find it more plausible that if safe, effective, and widely used genetic intelligence-enhancing technologies existed in rich countries, those countries would force their underclass to make use of them, rather than allow lack of resources to make the poor do without the technologies.

State governments in the United States spend on average over $10,000 per pupil for public elementary and secondary schools, with much of the money going to poor school districts.[316] According to that National Center for Education Statistics source, Washington, DC, where most public school students are from low-income families, spends $16,000 per pupil, but Andrew Coulson of the Cato Institute has made a strong case in the *Washington Post* that the District of Columbia actually spends closer to $25,000 per student.[317] The US federal government devotes about $9 billion annually to Head Start, an early childhood education program exclusively for children from impoverished families.[318] Politicians in the United States, we can conclude, have shown a strong preference for spending money educating the children of the poor. Should genetic enhancing technologies prove to be effective at increasing the intelligence of children, it's reasonable to assume that US state and federal governments will offer the technologies nearly free of charge to the poor.

If the affluent make wide-scale use of intelligence-enhancing genetics for their children, Americans across the political spectrum would almost certainly support governments making the services available to the poor. Conservatives who worry about the social pathologies of the underclass (many of which are correlated with low IQ) and dislike having governments spend money on welfare would support the poor receiving access to genetic enhancement technologies for their children.

Liberals profess a desire for equality of outcome, whereas conservatives claim to believe in equality of opportunity—both of which would become impossible if only the rich, or everyone but the poor, used genetic enhancements to create smarter children.

Recall from Chapter 7 that a nation's average IQ has a big impact on its growth rate. Because of political correctness, this fact is frequently ignored. But I predict that if genetic technologies make IQ much more manifestly malleable, many people will argue for raising the IQ of the poor in order to boost the national economic growth rate. If intelligence enhancements come not from drugs or genetics but rather occur through a Kurzweilian merger in which man and machine increasingly join, all of the same forces that push for equality will still hold.

Rich countries give billions of dollars in foreign aid each year to poor nations, and it seems reasonable to suppose that once cognitive-enhancing technologies prove themselves, part of this foreign aid will consist of giving these technologies to the world's most impoverished. Furthermore, it will be in the long-term self-interest of rich nations to help the people of poor countries boost their intelligence.

Rich nations gain enormous wealth through trade. All else being equal, a country gets more from trading with rich nations than it does from trading with poor ones because the richer you are, the better you are at creating stuff that other people want and the more you can afford to pay for stuff other people produce. Consequently, if it's believed that commonplace use of intelligence-enhancing technologies in poor nations would greatly improve these countries' economic growth, then advocates for international development aid will be able to truthfully claim that helping the world's poorest become smarter and have smarter children will, over the course of several decades, end up making everyone in the world richer.

Concern for racial equality will also motivate the United States and European countries to see to it that the world's poor are not left behind in the use of intelligence enhancements. For historical, colonial, racist, and environmental reasons, African Americans are the poorest major population group in the United States, and sub-Saharan Africa is the poorest inhabited region on Earth. If, therefore, the poor were excluded from the benefits of intelligence enhancement, then people with black skin would on average end up being much less intelligent than people with white skin, a prospect that would so horrify liberals that providing intelligence enhancements to the poor would likely become their top priority.

The poor will not be early adopters of expensive intelligence-enhancement technologies, since before the technologies become widely used, governments are unlikely to subsidize them. But once in wide-scale use, I'm confident that the poor will receive many of the benefits of these technologies.

It's even plausible that in rich countries, the genetic enhancements that poor children receive will do more to increase their intelligence than will the enhancements used on the children of the rich. Recall from Chapter 9 that with embryo selection, parents will probably face trade-offs among different favorable traits they might give their children. It's likely that most parents, if given the freedom to make whichever trade-offs they wished, would often be willing to give their children a bit less intelligence in return for the kids having some other positive traits, such as being good-looking. If, however, governments pay for enhancing technologies for poor children, it's plausible that they would prohibit parents from making superficial trade-offs and governments might therefore force fertility clinics serving the poor to give more weight to intelligence than affluent parents would.

The most likely exception to the equal distribution of intelligence enhancing-technologies arises if, as discussed in Chapter 14, the technologies turn out to be too risky for use by people in rich countries, but not too dangerous for nations such as China to subsidize or mandate.

If you can look into the seeds of time,
And say which grain will grow and which will not,
Speak then to me

—Shakespeare, Macbeth, Act I, Scene 3

CHAPTER 16
PREPARING FOR THE
SINGULARITY

While walking down the street, you feel a cold hand on your shoulder and turn around to see a strange-looking man touching you. He mumbles, "Here is your destiny" and hands you a scroll that contains fifty predictions for the next year and one prophecy that will supposedly unfold in 2045.

A year later you realize that every single one of the scroll's predictions for the year has come true. The accuracy of the scroll convinces you that its 2045 prophecy will also come to pass. The prophecy reads:

The year 2045 will bring the world one of three possible fates:
    1. Total annihilation—everyone dies.
    2. Immense wealth—everyone becomes rich.
    3. Unimaginable change—everyone's past wealth and material possessions no longer have value.

Since you're only thirty years old, you will probably live long enough to see what happens. The year 2045, however, is far enough away that you don't think the prophecy should affect your current behavior. But then, while shopping, you have a revelation. Your parents raised you to be financially prudent, and each year you put $10,000 aside in a retirement account for a retirement that won't start until after 2045.

But you now realize that if the prophecy holds true, you will never see any benefit from your retirement savings. If everyone dies, money means nothing. If everyone becomes rich, any money you saved in the past would be trivial—after all, if you were given $1 trillion in 2045, $500,000 in savings wouldn't really do much for your happiness. And obviously, if the world becomes so weird that money no longer has value, then you would receive no value from money saved.

When you save, you sacrifice your present for the future. While young and healthy, your parents bought less stuff than they otherwise would have so that they could buy more stuff when they were old and frail. By saving, they effectively transferred wealth from their young selves to their future, older selves. But because of the prophecy, you don't think that your younger self can make your older, retired self better off, and you decide to immediately spend the money in your retirement fund on gadgets and travel.

But then you rethink your decision. You can't be certain that the prophecy will happen. The scroll's fifty correct predictions provide powerful evidence, but not absolute proof, of the accuracy of the prophecy.

Savers gamble on the future. The chance that the prophecy will be right makes the odds on such a gamble less favorable, so you should save less for retirement than you did before reading the scroll but still put some funds aside.

After resolving to save $3,000 a year for retirement, you notice a blue light emanating from the scroll. You open it, and see additional text that reads:

> *You have used me well in making your decision, so I now reward you by revealing the rest of the prophecy:*
>
> *Almost Immortality—If the fate of either Immense Wealth or Unimaginable Change manifests, you shall live for millions of years.*

The new text causes you to sell your motorcycle, stop eating junk food, and quit your high-stress job for a more relaxed one. Before reading this final part of the prophecy, you figured that the cost of death was the fifty or so years of life that you would otherwise have experienced. But now if you die before 2045, you might miss out on millions of years of life. The high cost of death has made survival to 2045 your top priority.

This Book as Your Scroll

As you might have figured out, I'm the strange-looking man, and this book is your scroll. If you believe the arguments in this book, and think you have a good chance of surviving to the Singularity, then your expectations should influence your behavior today. And if/when other people come to believe in a Singularity, their behavior, too, will radically change.

SIGNPOSTS OF THE SINGULARITY

Several of the possible paths to the Singularity have signposts heralding their destination. For example, let's say that within fifteen years, someone makes a computer simulation of a chimpanzee's brain, puts this computer brain in a robot body, and the artificial chimpanzee acts just like a real chimp.[319] Many technology opinion leaders would understand that if we can create a computer chimp, we can create a computer human. And once we have a simulation of a human brain, we should eventually be able to increase the speed of this simulation a millionfold, make a million copies of the simulation, and usher in the Singularity.

Or perhaps within fifteen years it will be apparent to all who are technologically literate that within another decade an AI will pass what's known as the "Turing test," in which a human judge engaged in natural-language written conversation with the AI can't tell whether the AI is man or machine. And once this test is passed, we could eventually speed up the AI a millionfold, make a million copies of the computer, and produce a Singularity. Ray Kurzweil has bet $20,000 that a computer will pass a Turing test by 2029.[320]

Or maybe within twenty years, brain/computer interfaces will be developing at such a rate that an intelligence explosion seems inevitable.

Or improvements in gene therapy and eugenics could create millions of babies who, when they grow up, will be smarter than John von Neumann, and an understanding of what these babies will eventually accomplish could convince many that a Singularity is almost inevitable.

If any of these scenarios comes to pass, opinion leaders will come to expect that the Singularity is near, and this expectation will likely spread to the general population.

Politicians in debt-ridden countries such as the United States will have powerful incentives to convince voters that the Singularity is near. Because of their aging populations, governments in many rich countries will be spending a continually increasing percentage of their budgets taking care of the elderly. These demographic woes will force politicians to (1) significantly raise taxes, and not just on the rich, (2) cut other spending, and/or (3) run large deficits. Politicians hate the first two options because they upset so many voters. So far option (3) has been the favorite of governments, but this, too, concerns voters who object to piling debt on future generations. Politicians, however, might get away with borrowing huge sums of money if they could convince voters that intelligence-enhancing technologies will soon make everyone so rich that in the future paying off pre-Singularity debts will be trivial.

Knowing how mere expectations of the Singularity will impact society makes this book of higher practical value to you because it reduces the time before you can profit from understanding the Singularity. To identify profit opportunities, answer the following three questions:

1. How should I change my behavior because of my expectations about the Singularity?

2. How will others change their behavior when they come to expect the Singularity?

3. How can I profit from successfully predicting (2)?

I hope this book gives you guidance in answering these questions, but the most profitable answers will come from your specialized knowledge. Many readers have an extensive understanding of at least one sector of the economy. By figuring out how Singularity expectations will alter this sector, you will get financial information that almost no one else has.

DON'T DIE AND MISS OUT ON YOUR CHANCE OF LIVING A REALLY LONG TIME

I bet most of you reading this book are certain of your own eventual deaths. The universe has billions of years of life left in it, so even if you survived to age 150, you would have existed for a relative instant in time. On occasion, I suspect, most of you have been terrified by thoughts of the eternal nothingness that death might bring. And if you have children, the belief that someday they must die has probably brought you to tears. But if this book's thesis is correct and you survive to the Singularity, and the Singularity is good, then you will have the option of living a very long time.

In a Kurzweilian merger in which the Singularity comes through gradual improvement, the technologies to radically extend life will probably exist at least ten years before the Singularity. If we have a Kurzweil-type Singularity in 2045, then it's reasonable to assume that anyone who survives to 2035 and has access to the medical technology of rich nations will (barring accident) live to 2045. Life expectancy in the United States is currently seventy-eight years, and if you live this long, you will make it to 2035 if you were born after 1962. Kurzweil, however, was born in 1948 and has had diabetes, which shortens one's expected life span.

Kurzweil is smart, rich, and energetic, and thinks the Singularity will probably be utopian. If he meekly accepted a fate of not making it to the Singularity, I wouldn't trust that, deep down, he believed in the Singularity. But Kurzweil has extensively examined the scientific research on health, and he has adopted a diet and nutritional-supplement regime which he believes will maximize his chance of making it to the Singularity. He's even coauthored two health advice books titled *Fantastic Voyage: Live Long Enough to Live Forever*, and *Transcend: Nine Steps to Living Well Forever*.

The most important actionable implication of my book is that if you believe a friendly Singularity might be near, you should take extreme care with your health and safety to optimize your chances of making it to utopia. I have personally wondered how far to take this. Should I, for example, ever drive to a movie, knowing that an accident might deprive me of an extremely long life?

If you're male and want to do everything to maximize your chance of surviving to the Singularity, then, after verifying its veracity, act on the information in the following paragraph, taken from an article in *Scientific American* titled "Why Women Live Longer:"[321]

> A number of years ago castration of men in institutions for the mentally disturbed was surprisingly commonplace. In one study of several hundred men at an unnamed institution in Kansas, the castrated men were found to live on average 14 years longer than their uncastrated fellows.

To the best of my knowledge, no Singularitarian, not even Ray Kurzweil or bullet-eater Robin Hanson, is following the castration path to long life.

WHAT HAPPENS WHEN MANY THINK IMMORTALITY IS NEAR?

Businesses selling safety-enhancing products will be big winners when Singularity expectations cause many to become more fearful of death.

Traffic accidents kill over a million people each year, and are a leading cause of death among the children of the affluent. After the perceived cost of death rises, people will significantly spend more money on safer cars. Many Singularity predictors will be unwilling to work at dangerous jobs. Today, construction workers receive particularly high wages due to the hazards of their occupation. But once construction workers expect a Singularity, they will find this risk unacceptable, or will only take it on for much higher wages. Therefore, Singularity expectations will help companies that reduce the dangers of construction work.

Dreams of immortality will increase the demand for health care. I predict that Singularity expectations will cause voters to push politicians to spend a greater share of government revenues on health care. Many senior citizens, horrified at the prospect of just missing a utopian Singularity, might become single-issue voters and support whichever candidate offers them the best hope of survival. Understanding the stakes, non-seniors, I suspect, will become more tolerant of governments taking resources from them for health care for the elderly.

As voters, Singularity predictors will also want their governments to spend titanic sums on national defense. Precursor Singularity technologies might create tremendous military uncertainty by destabilizing the existing balance of power. When you combine military uncertainty with citizens' extreme fear of death, the political demand for military spending will be high. And, of course, the more one nation spends on its military, the more its adversaries will want to spend as well. So when the Singularity is widely expected, we might see a massive increase in worldwide military spending. Making smart investments today could allow you to profit from this increase.

Discounting the future will cushion the impact of immortalist expectations. If given a choice between receiving $100,000 today or $200,000 in twenty years, many people would take the immediate cash because they do what economists call "discounting." Social scientists observe that most people place much less importance on what will happen in the far future, compared to what's going to occur this year. If, for example, you greatly discount the future, then you won't save for retirement. And smoking becomes rational because it brings immediate benefits, whereas most of the costs won't hit you for at least a few decades.

If you highly discount the future, then living another million and fifty years wouldn't mean much more to you than living another fifty years because you care very little about what will happen to you several decades hence. Most preteens—people challenged to worry about anything that won't transpire within a year—almost certainly wouldn't act differently if their expected life span increased by a million years. Substantial discounting would also cause some adults not to be affected by learning that their expected life span has increased manyfold.

I doubt, however, that many adults, upon learning that they have a decent chance of surviving for millions of years, would discount the future so much that expectation of an extremely long life wouldn't radically change their behavior. People discount the far future partially because they think that when they are elderly, ill health has a high chance of preventing them from reaping much enjoyment from life. In an extreme case, if you knew that you would get

Alzheimer's disease when you were eighty, you wouldn't worry much about staying healthy today in order to still be alive then. Most people, though, would likely be much happier in a utopian Singularity than they are today. Expecting the Singularity would cause many to start looking forward to the far future and to be more concerned about dying today than they otherwise would.

BUSINESS INVESTMENTS

Many businesses make investments on which they don't expect to break even for a long time. Singularity expectations, or even expectations of rapid technological development, would kill many of these investment projects. For example, consider a real estate developer deciding whether to construct a large office building that will be rented out to companies over the building's sixty-year life. This developer would have many reasons to think that technology might decrease the value of the building. Clearly, a full-blown Singularity that destroys the value of money would stop the office building from yielding revenue to the developer. But also, if robots become able to construct buildings cheaply, they will increase the supply of buildings and lower the price of office space. If, through nanotechnology, it becomes trivial to construct large office buildings, then the value of the building will fall to the worth of the land it sits on. The developer might also fear that widespread use of virtual reality would lower the demand for real-world office space. Even if a Hanson emulation scenario makes the land fantastically valuable, having a building on the land wouldn't increase the land's value, because in a world of trillions of emulations, the building would likely be razed and turned into a super-skyscraper to accommodate the digital hordes. The possibility of these technologies will reduce the developer's estimate of future profits and so decrease the chance that she will construct the building in the first place.

Developers of residential housing might also scale back activities due to Singularity expectations. The house I currently live in was built in the early 1970s, although my wife and I purchased it in 2004. The original buyers of the house no doubt believed that the home would retain considerable value and expected that they would recoup at least most of the home's purchase price when they eventually sold the house. When deciding how much they're willing to pay for a house, potential home buyers often take into account the home's resale value because purchasing a home is a combination of consumption and investment. Several of the homes my wife and I looked at had only one story, and I remember my wife telling me that elderly people much prefer homes that don't have an upstairs. Since the elderly population of the United States is growing, such houses are likely to increase in value more than comparable two-story homes will.

The price of lived-in homes can't get too much higher than the price of comparable new homes. Consequently, the resale price of existing homes won't surpass the total cost of building a new home. Therefore, home buyers who expect that new technologies will reduce the costs of building new homes will have a lower estimate of their homes' future resale value and will pay less for housing today. This falling demand will cause real estate developers to build fewer homes, or at least to build cheaper ones. All of this happens not because anyone has yet developed Singularity technologies, but because people expect that these new technologies to come online within three decades' time.

Creators of media content might also make fewer investments based on an expectation that artificial intelligence would lower the cost of producing goods in their market. Owners of copyrights on movies, music, and books receive revenues on content created decades ago. When deciding whether to fund new projects, media companies take into account all the revenues the content will ever bring in. Artificial intelligence, however, might become extremely good at creating content. Perhaps you will be able to tell an AI your favorite songs, and the AI will then create customized music that you'll enjoy so much that you'll never again purchase music produced by humans. Or a media company might think that cognitive-enhancement drugs will soon improve the skills of authors, so that novels written this decade will effectively become obsolete, as all readers will find them inferior compared to novels crafted by newer authors. Even if a publisher believes there is only, say, a 20 percent chance that technology will render their backlist obsolete, this 20 percent expectation will reduce the advances publishers pay authors, which will in turn reduce the number of novels that authors pen.

Energy companies might also make fewer investments if they start to expect that artificial intelligence will come to play a big role in their industry. Ray Kurzweil has written that precursor Singularity technologies will do much to create and conserve energy.[322] A utility firm considering a huge long-term investment in an electrical plant might be dissuaded if it believes that the technologies that Kurzweil describes will reduce the demand for electricity or produce it far more cheaply.

Singularity expectations will also reduce the willingness of companies to create new products for which they would expect to generate sales for long periods of time. An automobile manufacturer, for example, might become unwilling to spend five years developing a new type of car if it expects that in ten years AI-driven cars will render its new design obsolete.

Singularity expectations, however, won't reduce investments in all types of goods. The shorter the expected life span of the product, the fewer investment decisions will be inhibited by the expectation of rapid technological improvements. If you're going to develop a video game that will generate almost all of its sales within a year, you wouldn't care if, in two years, virtual-reality technology makes the game you're working on seem pathetic.

WAYS EXPECTATIONS COULD INCREASE RESEARCH AND DEVELOPMENT

Expecting artificial intelligences to greatly speed up drug development could cause pharmaceutical firms to spend more on research today. Let's consider a simplified two-stage model of drug development and pretend that to get a new drug to market a pharmaceutical company will first spend seven years sifting through possible chemical compounds to find a few that might have medical value. After finding some candidate compounds, a drug company then spends seven years testing the compounds on animals and people to verify the drugs' safety and efficacy. If the pharmaceutical company thinks that artificial intelligence would soon so vastly increase our ability to find and test new compounds that any drug that got started now would never reach the market (because it would be inferior to something that future AIs would make), then the firm would spend nothing on new drug development.

But a pharmaceutical company might come to think that artificial intelligence would be much better at helping with Stage II than Stage I of the drug development process. Perhaps the company develops drugs to treat depression and thinks that in seven years computer models of the human brain will make it possible to test a virtual version of the drug on a simulation. Successful virtual testing might reduce the number of animal and human tests needed to get the drug to market. Today, the pharmaceutical company might start developing the antidepressant drug just because it predicts that the final cost of developing this drug will be much lower than it would have been without the not-yet-developed AI technology.

Think of pharmaceutical product development as going on a journey that will take fourteen years in your current vehicle but will require less than seven years in a vehicle that you think will be available for use seven years from now. You can either start the journey today and then switch vehicles in seven years, or wait seven years and complete the entire journey in the new vehicle. If you had to start the journey over again when you got your new vehicle, you would always choose the latter course.

Let's now suppose that in seven years you will be able to get the vehicle delivered to the position you have reached so far. Then you'll be better off in seven years for having started the journey today. If the revenues you will earn from completing the journey would be less than the cost incurred in taking a fourteen-year journey, then you might start the journey today only because of your expectation of receiving the new vehicle. Of course, if the new vehicle will be sufficiently fast, you would still wait to receive it before beginning a trip.

Expecting better technologies will spur companies to spend more on "treasure hunt" investments. Pretend that lots of people have copies of old pirate maps that are written in code and show the location of a buried treasure. You win the treasure if you're the first to find it. The maps, however, clearly state that the treasure fell through a sinkhole, and everyone realizes that with today's technology, they wouldn't be able to extract the treasure even if they knew its location. If, however, map holders expected that future improvements in technology would make it possible to dig up the treasure, then they might immediately start to look for the treasure so they could buy up the land under which it lies and have first dibs on it when the technology to extract the treasure arrives.

Energy exploration companies often do "treasure hunts" in which they seek out the location of fossil fuels. These companies might spend more today on seeking out fossil fuels if they believed that within the next decade artificial intelligence will greatly lower the cost of extracting energy from the earth.

Pharmaceutical companies also engage in a bit of treasure hunting. These companies receive patents, which give them the exclusive right (for a period of time) to sell the drugs they have developed. Essentially, each compound is a treasure to which the patent office assigns an owner. Many drugs provide several medical benefits, not all of which are known at the time of the drug's discovery. For example, modafinil was created to treat narcolepsy but is now widely used as a cognitive-enhancement drug by non-narcoleptics. If a drug company expects that in the future, artificial intelligences will find different uses for already-patented drugs, the company might devote additional resources to drug development today because it believes that owning drug patents will become more valuable in a future in which these artificial intelligences exist.

FINANCIAL INSTITUTIONS

One potential objection to my Singularity expectations argument is that most people lack even a rudimentary understanding of genetics, exponential trends, neuroscience, and computational science and are extremely unlikely to ever realize that the Singularity is near. I believe that many will come to expect the Singularity, but even if I'm wrong, Singularity expectations will still have a massive impact on the economy if a few thousand hedge fund managers and venture capitalists become Singularity predictors.

Hedge funds take money from rich people and invest in financial instruments such as stocks and bonds. These funds collectively control trillions of dollars in assets. In 1998, the hedge fund Long Term Capital Management had, all by itself, financial positions of around $1.3 trillion.[323]

Because of the huge sums involved, hedge funds attract the best and brightest as employees. Consider a hedge fund that has convinced rich people to invest $100 billion with it. The hedge fund then uses this $100 billion to borrow $900 billion from banks, allowing the fund to make financial bets totaling $1 trillion. Pretend that if this fund were run by the second-most-qualified person on Earth, it would earn a 10 percent yearly return, but if the fund were managed by the best person, it would have a 10.2 percent return. This difference amounts to $2 billion a year, meaning that the fund would greatly benefit from employing the best manager for a salary of $1 billion a year, a salary that would tempt almost anyone.

Hedge funds do sometimes screw up. Long Term Capital Management, for example, lost billions in 1998, and the US government so feared what would happen to the world economy if the fund imploded that it arranged a bailout. The ever-present possibility of failure motivates hedge funds to eagerly seek out all relevant information. Hence, if a very smart, rational, and well-informed person should be able to see a signpost to the Singularity, then you can bet that several hedge fund managers will read and act on such a sign.

If hedge fund managers make investment decisions based on their expectations of the Singularity, then many asset prices will come to reflect those expectations. Pretend that a hedge fund believes that the correct value of a share of Intel stock is $50. If the price of Intel goes below $50, the fund will buy the stock, raising its price to $50. If the price of Intel goes above $50, the hedge fund will sell Intel, pushing the price down to $50. Hedge funds that find mispriced financial assets have incentives to buy or sell the asset until its price becomes equal to what the fund thinks it should be. A financial asset will be mispriced if the asset's value doesn't reflect all available information, such as the likelihood of the technological Singularity.

But, you might ask, "So what?" Why does the price of financial assets matter? It matters because the higher a company's stock price is, the better off its investors are. Most companies put a lot of effort into raising the value of their stock. So here is how Singularity expectations will influence corporate behavior:

1. Hedge fund managers have the incentive and intelligence to spot Singularity signposts.
2. Hedge fund managers have a huge amount of influence over stock prices.
3. Companies care a lot about stock prices.
4. Companies, therefore, have a strong incentive to make decisions that take into account the possibility of a Singularity.

Now let's turn from hedge fund managers to venture capitalists. Venture capitalists raise money to invest in new, often technology-based firms. A venture capitalist takes equity in new firms in return for funding them and providing them with guidance. Obviously, the better the firm does, the greater the venture capitalists' profits.

Lots of smart, hard-working innovators have ideas they want to turn into businesses; they dream of creating the next Facebook or Google. Most of these innovators require financial backing, and they often get this backing from venture capitalists. Venture capitalists can't and certainly don't want to fund every business idea, so they often act as gatekeepers to innovators. For an innovator to open this gate, he must please the venture capitalist.

Venture capitalists who invest in technology have a huge incentive to correctly predict how technological developments will unfold. If the arguments in this book are correct, therefore, venture capitalists will soon recognize the signposts to the Singularity. And if venture capitalists expect the Singularity, then innovators hoping to get funding for a new business will be forced to consider how their business's profits will be affected by the Singularity. Peter Thiel, a key financial backer of the Singularity Institute for Artificial Intelligence, is both a hedge fund manager and a venture capitalist. Many venture capitalists have attended conferences at the Singularity Institute. While I was writing this book, a hedge fund manager e-mailed me asking about a short article I had written on how intelligent robots would affect the wages of human workers. This manager informed me that he had talked to Peter about this issue.[324]

SINGULARITY EXPECTATIONS AND SAVINGS

How Singularity expectations will influence savings rates (the percentage of income saved) depends on how people will answer these questions about the type of Singularity they predict:

1. Will your pre-Singularity property rights be respected post-Singularity?

2. If property rights will be respected, what return will you receive on pre-Singularity investments?

3. Will you be wealthy post-Singularity?

SAVINGS AND PROPERTY RIGHTS

If you believe that any savings or investments you own pre-Singularity will be worthless post-Singularity, then you won't bother to save for the long term. An ultra-AI might obliterate the value of pre-Singularity investments by taking control over the distribution of resources and allocating them without regard to pre-Singularity property rights. A Singularity might also abolish scarcity, change humans so that we no longer value wealth, or do something else that I haven't considered that eliminates the value of pre-Singularity investments.

A technological Singularity would also bring about rapid political change. In the past, such change often destroyed old property rights. After their conquest of England in 1066, the Normans quickly appropriated much of the wealth of the old English aristocracy. The Aztec rulers lost their wealth to Cortez. Following the North's victory in the US Civil War, the United States abolished slavery without paying compensation to the slaveholders. The Nazis destroyed the value of German Jews' investments through theft and murder. And the communist revolutions in China, Russia, and North Korea not only eliminated the value of pre-revolutionary industrial investments but also gave those investments a negative value because individuals who had significant wealth pre-revolution were considered enemies of the state.

Recall that in Chapter 3 I wrote that we should be terrified of an unfriendly AI, in part because an ultra-AI could completely rearrange the distribution of atoms in our solar system, and only in a minuscule percentage of these distributions could mankind survive. The percentage of distributions in which current property rights are respected is even smaller.

As discussed in Chapter 13, economist Robin Hanson believes there is a significant chance that pre-Singularity property rights would still be respected after a Singularity. Robin believes it unlikely that an ultra-AI would emerge from an intelligence explosion having the capacity, let alone the desire, to extinguish mankind. Robin also thinks that the most likely path to a Singularity is through emulations, and that if bio-humans don't attempt to control more than, say, 5 percent of a society's wealth, then emulations probably won't team up to take all of our stuff, even if they outnumber us by more than a trillion to one.

In what I'll now call a Ricardian comparative advantage Singularity (see Chapter 13 for why I picked this name), in which a few AIs emerge that respect property rights and trade with us, then (by definition) pre-Singularity property will have value post-Singularity. In a Kurzweilian Singularity in which man and machine steadily merge, property rights will also likely be respected post-Singularity, although you could argue that once the concentration of intelligence in the solar system becomes high enough, civilization will be so much different from today's that it's impossible to make reasonable forecasts about property rights.

SAVINGS AND INVESTMENT RETURNS

Overall, the faster an economy grows, the greater the benefit you receive from investing in it. The many business opportunities that a rapidly expanding economy creates raise the demand for investor dollars, and so businesses competing for investors must offer large expected returns. Under either Robin's emulation scenario or Kurzweil's merger of man and machine, the economy will experience extremely fast growth, creating high average returns on investments, which (all else being equal) will raise savings rates. For example, if you knew that your bank would triple the value of everything in your savings account next year, then you would likely save vastly more so that you could benefit from this bonanza.

The expected obsolescence of many types of investments, however, such as investment in real estate development, would diminish savings. You would be reluctant to put a lot of money in the stock market if you expected that most of the companies you invested in would have their value destroyed by new technologies. The general rule in economics is that (other things being held constant) the more abundant something is, the lower its market value, and the more scarce a resource is, the greater its price. This holds true for labor and capital. If cheap emulations cause there to be far more workers than "production wands" (land, machines, and materials you need to create goods), then production wands will greatly increase in value. Also, to the extent that pre-Singularity people expect this, they will save much money so they can invest in production wands. In contrast, if we never have emulations, but AIs give us powerful nanotechnology allowing us to construct production wands cheaply, then these production wands will have relatively little value. Those expecting this to happen will intentionally save little.

Consequently, the rate of return on savings will be high in Hanson's emulation scenario because there will be so many more workers than production wands. In contrast, under a Ricardian comparative advantage Singularity, there will not be a vast number of workers, but nanotechnology and other types of production technology will probably make it very cheap to construct new buildings, and rates of return on savings will be relatively low.

SAVINGS AND FUTURE WEALTH

If a squirrel were uplifted to human-level intelligence, assimilated into human society, and employed at a job that paid $50,000 a year, the squirrel wouldn't gain much benefit from the nuts it had stored for the winter. The uplifting would have essentially annihilated the value of the squirrel's investments by making him so wealthy that his stored nuts became worthless post-uplift. Similarly, a utopian Singularity might give humans enough resources so that anything anyone owned pre-Singularity became trivial. Anyone expecting this type of utopian Singularity would greatly curtail his savings.

Under Robin's emulation scenario, bio-humans have an excellent chance of receiving large welfare payments, and any pre-Singularity investments we have will turn out to have immense post-Singularity value. Emulations, however, would drive wages to near zero, which would effectively impoverish anyone who didn't receive welfare or have investments. In a Kurzweilian merger, humans would likely be fantastically productive and earn higher wages, ensuring that almost everyone would be very rich, and expecting this would reduce pre-Singularity savings. Under the Ricardian scenario, our wages will also be extremely high, although not as high as under a Kurzweilian merger because under Ricardo our bodies will not possess the productive power that they could acquire by merging with machines in a Kurzweillian scenario.

SAVINGS RATES WITH THE DIFFERENT SINGULARITY SCENARIOS

This chart shows how expectations of different types of Singularities will influence savings rates. It does not show what savings rates will actually be like if a Singularity occurs.

Type of Singularity	Property Rights Expectations	Rate of Return Expectations	Wealth Expectations	Cumulative Effect on Savings
Intelligence Explosion	Probably not respected—greatly lowers savings	Not relevant	Rich if not dead—greatly lowers savings	Lowest of all types of Singularity
Kurzweilian Merger	Very likely respected	Ambiguous—rapid economic growth raises returns, but cheap production wands likely lower them	High—lowers savings	Savings rates will likely be higher than if people expected an intelligence explosion, but less than if people expected emulations
Ricardian Comparative Advantage	Respected	Ambiguous—rapid economic growth raises returns, but cheap production wands likely lower them	High, but not as high as for intelligence explosion and Kurzweilian merger—lowers savings	Savings rates likely greater than if people expected an intelligence explosion, but less than if people expected emulations
Emulations	Probably respected	High—rapid economic growth and a large number of people relative to production wands will greatly increase savings	Ambiguous—If rely on wages then poor; if have welfare or if own production wands, then rich	Highest savings rates of the four types of Singularities

If Savings Rates Plummet

Since I think that the most likely type of Singularity will occur through an intelligence explosion, I shall now use some simple economic analysis to explain what would happen if savings rates plummeted.

When you save, your money goes from you to a borrower, either directly or via a financial intermediary such as a bank. Many borrowers invest the money they receive from savers, and these investments help to grow the economy. Most of the funds companies use to engage in research and development ultimately come from household savings. Because research and development will likely play a big role in determining when we achieve a Singularity, the big conclusion of this subsection is that:

> *Expectations of the Singularity have a reasonable chance of postponing the occurrence of the Singularity.*

Think of the world's economy as a giant farm that produces food every year, some of which the farmers eat and the rest of which they sell for money, which they use to purchase seeds that will be planted for the next harvest. The farmers are always hungry, and therefore selling food entails making a short-term sacrifice. For these farmers, saving less means eating more food today and consequently having fewer seeds to plant for next year.

Now pretend that once the total amount of food produced exceeds some given amount, there will be a Singularity that will cause farmers to have as much food as they desire forever after. If a farmer knew that the coming harvest would bring such a Singularity, he wouldn't trade any of his food for seed. But let's suppose that none of the farmers know exactly when the Singularity will arrive, but they think it's coming soon. The more likely a farmer thinks it is that the Singularity will arrive with the next harvest, the less food he will forgo eating. Since there are so many farmers, no single farmer's decision more than trivially affects when the Singularity will transpire. Furthermore, because the farmers don't coordinate their decisions, no farmer thinks that his saving less would cause other farmers to also reduce savings. Consequently, many farmers thinking that the Singularity is near will, paradoxically, result in less food being produced at the next harvest, and so reduce the likelihood that the Singularity will happen at the next harvest.

Here's another analogy for how expectations of the Singularity might push back the date of a Singularity: imagine society is building a Singularity machine that (when finished) will create a Singularity, and so destroy the value of money. The machine requires the efforts of many individuals, and the government induces them to work by giving them money, but the money can't be spent immediately. As the machine draws closer to completion, it becomes harder and harder to pay workers because the money they receive has less value to them. Therefore, when the Singularity is perceived to be near, the government can't afford to hire as many workers, which increases the time it takes to achieve a Singularity.

EDUCATION

Education is an investment in oneself, and consequently will be impacted by Singularity expectations. I teach at Smith College, an expensive private school. Its students expect to recoup the cost of their education over their working careers. But if the Singularity appears near, students will correctly figure that the chance of their education still being useful to them in fifteen years will be low.

The value of education post-Singularity will depend on the path to Singularity that mankind takes. Obviously, if the Singularity destroys us, education will have no post-Singularity value. Similarly, if the Singularity makes us all extremely rich or changes the world in a strange enough way that money has no value, than being well-educated will not make you materially better off.

In a world of cheap emulations, most people's education would yield them no financial value, since wage rates would be near zero. But if your education causes you to be one of a few to be emulated, then your biological self could earn high profits by selling the copyright to your brain template. If emulating new brains becomes inexpensive, then you would also derive no monetary gain from your education even if your emulations benefitted from your education. The ability to cheaply emulate bio-humans would also cause prices for the right to emulate someone to be near zero, since there would undoubtedly be many people who would relish the chance to have themselves emulated, and so would freely give away their brain template rights.

Being educated would probably offer the greatest payoff if we achieve Singularity through the merger of man and machine, since under this Kurzweilian scenario the machines in your brain would complement your normal education. Economists like myself could have a math coprocessor implanted in our brains. If, however, brain implants would make it trivial to download knowledge and become experts on anything, then past education would not confer any extra economic value.

Before the Singularity, below-human-level AIs will raise the value of certain types of education but lower the value of others. The financial value of being educated has been steadily going up as nations become richer and more technologically sophisticated, and we should expect that better technology will (on average) raise the wages of the well-educated.

Below-human-level AIs will almost certainly make some skills, such as proficiency (but not fluency) in a foreign language, of minimal economic value. Because of the rapid growth of the Chinese economy, many people think that Mandarin is the language of the future. But American parents of young children probably shouldn't pay for Chinese lessons to improve their children's long-run economic prospects, since by 2025 there will almost certainly be wearable computers that automatically translate anything you hear in Chinese into English and anything you say from English into Chinese. Of course, if the child enjoys learning the language, or if learning a foreign language helps a child develop intellectual discipline, then a child could gain from taking Chinese lessons.

REPRODUCTIVE TIMING

Besides affecting what you teach your children, Singularity expectations should affect when you have children. A belief that future improvements in genetic technology would increase the "quality" of children would cause some women to postpone reproduction. Also, if Singularity expectations significantly raised people's uncertainty about the quality of the future, many might wait to have a baby until they learn how the Singularity turns out. By reducing the nonworking percentage of the population, child postponement would boost economic activity, perhaps even speeding up the arrival of the Singularity.

EXPECTING SEXBOTS

Dilbert creator and blog writer Scott Adams wrote, "I asked how many guys would have sex with a robot if it was indistinguishable from a hot human woman. About 95 percent of the hetero guys said they would. The other 5 percent expressed a strong preference for lying."[325] In light of the divorce statistics in the United States, sexbots would especially appeal to men concerned about the financial devastation that divorce can bring.

My guess is that we will have robots that look exactly like people before we will have robots that act just as people do. Consequently, the more weight you give to appearance when picking a mate, the sooner you will find sexbots to be a superior substitute to real people. Since, on average, men are more superficial in their sexual desires than women, initially sexbots will take more men than women out of the dating pool.

Cheap sexbots would harm an economy by causing men to put less effort into their careers. Economist Eric Gould has estimated that "if there were no returns to career choices in the marriage market, men would tend to work less, study less, and choose blue-collar jobs over white-collar jobs."[326] A government, however, could combat this negative economic incentive effect by imposing a higher tax on sexbots based on the higher the "quality" of the sexbot, so men would still have an incentive to work hard to acquire a high-quality "mate."

Widespread sexually transmitted diseases would increase sexbot use.[327] I can imagine the government of a country racked by an epidemic of STDs giving away sexbots. A nation might also employ this strategy to reduce teen pregnancies.

The downside of sexbots for men is that, at least for a while, they won't be able to produce children. Men who want children and sexbots could, however, still raise a child with a woman, or hire a surrogate to help produce a kid. Sexbots would likely increase utilization of eugenics technology by further separating reproduction from mating, making sexbots particularly attractive to pro-eugenics nations like China.

I predict that the Chinese government will enthusiastically embrace sexbots. Besides the eugenics benefits, sexbots would solve China's woman shortage problem. China's one-child policy combined with its citizens' preferences for having at least one son has caused many Chinese parents to seek sex-selective abortions. A 2005 population survey estimated that among people under twenty, there were thirty-two million more men than women.[328] A desire to limit population growth would probably provide another reason for the Chinese government to promote sexbots.

Opposition to sexbot use will undoubtedly arise. Nations such as Russia and Japan that are concerned about underpopulation will fear the sexbot's negative effect on reproduction. Politicians might oppose sexbots out of puritanical concerns, although such concerns have proven impotent in restricting Internet pornography. Feminist groups might oppose sexbots because by reducing the percentage of men in the dating pool, sexbots would increase the dating market power of the men still interested in dating humans, and consequently the sexbots would turn the dating market "terms of trade" against women. Lesbian feminists, however, would have less objection to sexbots. Some feminists might actually support sexbots if they believe that widespread use of them would reduce rape.

The mere expectation of the future development of sexbots could have profound effects on the behavior of socially awkward young boys. Currently, many of these boys put lots of effort into figuring out how to relate to women because they hope to someday win a mate. But if these boys expect that sexbots will someday become excellent substitutes for women, they might stop trying to develop the social skills needed to attract females.

The best evidence for the future impact of sexbots comes from how pornography is already impacting the sex drives of today's twentysomethings. As *Psychology Today* reports, porn is causing a significant number of young, healthy men to no longer be turned on by their real girlfriends.[329] Just imagine what will happen (and soon lots and lots of men will be imagining this) when porn becomes "touchable," three-dimensional, and fully interactive. I predict that for many people, sexbots will be to desire as candy is to hunger.

CHAPTER 17
WHAT MIGHT DERAIL THE SINGULARITY?

Predictions live in the realm of probabilities, not certitudes. Although I believe that increases in human and artificial intelligence will probably play an enormous role in shaping the next sixty years, I and other well-informed commentators have doubts. Since I'm trying to be more of a judge than an advocate with this book (except for a chunk of material in the next chapter), I now examine four categories of arguments against the likelihood of radical intelligence enhancements.

1. Civilization Collapses

In my opinion, the most probable reason why mankind will never experience significant increases in machine or human intelligence is that our high-tech civilization won't survive long enough for it to happen. That is, nuclear war, biological or nanotech weapons, or natural disasters such as super-volcanoes or asteroid strikes wipe out our species, or at least send us back to the Stone Age.

One of the most powerful, but strangest, arguments that civilization will probably soon collapse comes from Robin Hanson's application of what's known as Fermi's Paradox. To give you an intuitive grasp of the argument, I present the following story:[330]

One day you wake up with a strange kind of amnesia in which you have forgotten everyone's age and lost the ability to determine people's age from their appearance. You look at your driver's license and see that you are 67. Since you don't know how long members of your species usually survive, you can't estimate your remaining life span. To gauge the odds of making it past age 70, you walk around town asking everyone you meet if they are over 70. You never ask anyone if they are between 67 and 70. Finding many 70+ members of your species would mean that you have a decent shot at living many more years. But everyone you talk to says something like, "No, I'm younger than 70." Fear of rapidly-approaching death causes you to spend a year searching for someone 70+, but you find none.

Scared, you worry that if no one else ever makes it past 70, you will soon perish. But then you think of a few reasons why you might not be doomed. First, perhaps all 70+ people withdraw from society, possibly going into secret nursing homes. This, however, seems unlikely, because you are in excellent shape and can't imagine that you would confine yourself to a nursing home if you reached 70. Then, you consider the possibility that you might not have searched long enough. Conceivably, many 70+ individuals roam the streets, but you've been too unlucky to find them. This, however, seems statistically unlikely.

Then you hit upon a reason that makes you a tad optimistic: perhaps no one else has lived to 70 because they all died before they reached your current age. If everyone but you dies before, say, 60, then you would have already made it past life's main choke point and might well have many decades left. Your failure to find any 70+ people could, you hope, bode well for you because it might provide evidence that your body possesses some extraordinary strength and resilience.

But, you think, if you are the only person, or at least one of the very few people, to make it to your current age, then you're extremely special. In contrast, if many people die just short of their 70th birthday, then you're relatively ordinary age-wise. You realize that one should always be wary of theories based on the theorist being extraordinarily special—which is why ancient astronomers should have been suspicious of their models that assumed that they lived at the center of the universe.

You get a physical examination from a pediatrician who has never before treated an adult and who has never bothered to learn about the health or longevity of adults. The doctor says that while he can't find anything wrong with you, something in your genes will trigger a bodily change before you reach 70 that will either kill you or make you fantastically healthy, extending your life span by centuries. The doctor explains that the genetic change must happen to everyone before hitting 70.

You take the doctor's diagnosis as proof that you will soon die. If the change is always fatal, it explains why you haven't met any 70+ people. If the change bestowed longevity on many people, it would increase the number of 70+ people in society, making it even more statistically unlikely that you would not have found any.

FERMI'S PARADOX

Fermi's paradox makes mankind's short-term survival prospects analogous to yours in the previous story. Enrico Fermi, a physicist who helped the United States develop the atomic bomb, wondered why we haven't seen any evidence of intelligent extraterrestrial life. Our failure to find extraterrestrial life is similar to our protagonist not finding any 70+ people.

The Milky Way galaxy contains at least 100 billion stars, many of them older than our Sun. Traveling at only 1 percent of the speed of light, an advanced alien could cross our galaxy in ten million years, a tiny amount of time compared to how long life has thrived on Earth. Over a trillion stars lie within three million light years of Earth, a distance a post-Singularity civilization could manage in a (relatively) short period of time. And if the universe allows super-luminous travel, then vastly more stars could have spawned a civilization that would have had time to reach Earth.

If the universe contains a limited amount of free energy, then it's even more surprising that we don't see evidence of intelligent life. Since stars waste huge amounts of free energy, you would think that an advanced civilization would "turn off" stars to conserve energy, or at least capture as much energy from as many stars as it could. So the fact that we see shining stars decreases the chance that there are star-faring civilizations within a few million light-years of us.

Evolution, which must have shaped the objectives of any existing alien races, favors species that spread out and reproduce, and so, you would think, a significant percentage of extraterrestrial species would become star-faring as soon as they were technologically able. Evolution, therefore, strengthens Fermi's paradox.

A post-Singularity race could quickly spread throughout the galaxy using a type of spacecraft imagined by John von Neumann. Von Neumann proposed that, with sufficiently advanced technology, you could build a self-replicating probe that would land on planets or asteroids, make copies of itself, and send the copies out to find new planets where they could further replicate themselves. The probes could also perform useful tasks, such as sending back information or preventing stars from wasting free energy. A post-Singularity civilization could probably send out a single von Neumann probe at a trivial cost.

It's possible that all the arguments in this book are wrong, and that the laws of physics outlaw superhuman intelligence. But all you would need to make a von Neumann probe is strong nanotechnology, and given how much nature makes use of nanotechnology, it's hard to see how a civilization just a few hundred years in advance of us wouldn't have enough nanotech to build a von Neumann probe.

Intelligent alien life might be hiding from us. But given how many stars exist in Earth's neighborhood, if life isn't extraordinarily rare, it's hard to believe that thousands or millions of advanced life-forms would all unanimously keep us in a cone of ignorance.

Economist Robin Hanson has explained Fermi's paradox in terms of a "great filter."[331] Something seems to make it extremely difficult for star systems to give birth to star-faring species. Robin's great filter represents whatever forces prevent space-traveling civilizations from arising in almost all star systems. The critical question for humanity is whether our Solar System has already passed through the great filter or whether the great filter is something we will have to pass through in the future before we can leave our Solar System. If we have already gone through the filter, then, as my story's protagonist hoped, we have somehow escaped the near-certain death that killed off other systems' chance of creating a star-faring race. If, however, we have yet to face the great filter, then we are almost certainly doomed. Learning that we have never seen extraterrestrials because all types of life are extraordinarily rare in the universe would be much better news than learning that we have never encountered evidence of extraterrestrial intelligence because virtually every single civilization destroys itself within a millennium of discovering calculus.

Candidates for an early filter include the difficulty of simple replicators forming, the challenge of these replicators evolving into life complex enough to spawn a technological civilization, and the unlikelihood that any given planet would have an orbit that would allow for long-term stable climates.[332] If the filter comes after us, it must be due to hit relatively soon because we seem close to becoming a star-faring race. If (a) Kurzweil is right and we will undergo Singularity by around 2045, (b) we haven't yet escaped the great filter, and (c) the reasoning in this chapter is correct, then our likely extinction is so close that most readers of this book shouldn't bother saving for retirement.

Of course, the filter doesn't have to be at just one point in time. We might have passed through some but not all of it. Although a multi-filter scenario would give us better odds than if the filter just lies ahead of us, we still might have only a minuscule chance of survival.

Let's say that of the trillion star systems closest to us, one hundred have birthed star-faring civilizations. This seems to be approximately the largest number of such civilizations that could exist without making evidence of extra-terrestrial intelligence overwhelmingly obvious. If we have passed through only half of the filter, then our chances of survival are beyond tiny. But let's imagine that we have passed 99.9999999 percent of the filtering. How difficult, then, will it be to pass through the last .0000001 percent? Unfortunately, .0000001 percent of a trillion is 1,000, meaning that out of the 1,000 civilizations that have reached our level of development, only 10 percent survived much longer.

Robin became extremely pessimistic after reading an analysis by Katja Grace, former intern at the Singularity Institute, who used anthropic reasoning to estimate the temporal location of the great filter.[333] Anthropic reasoning involves drawing inferences from the fact that you exist. The story at the beginning of this chapter was my attempt to convey a few of Katja's insights.

Since we have a nontrivial chance of destroying ourselves with weapons of mass destruction, we almost certainly haven't passed all the way through the great filter. But if we have passed 99.9999999 percent or more of the way through, then we occupy an extremely special position. In contrast, if a fair amount of the great filter lies in front of us, then we occupy a "normal position" in the universe's history because given the exponentially increasing population growth that is likely to occur when a species approaches our level of development, many intelligent life-forms will find themselves alive when the great filter annihilates their civilization.

If the Singularity is near, then it becomes even more likely that we will soon perish because if civilizations at our level of development usually undergo a Singularity, then this Singularity becomes a plausible great filter, analogous to the genetic change that the protagonist in the previous story would undergo. Furthermore, as a friendly Singularity would make it easy for us to spread throughout the stars, the possibility of a Singularity further reduces the chances that an advanced alien civilization wouldn't yet have found us if such civilizations exist.

Many scientists, however, view anthropic reasoning with suspicion, and so perhaps we shouldn't use it to conclude that we will soon face doom. Still, anthropic reasoning should give us further cause to fear the Singularity—although, if much of the part of the great filter that we have not yet passed through springs from pre-Singularity destructive technologies such as biological weapons, then quickly achieving Singularity becomes our hope of escaping the great filter.

Anthropic reasoning, as Katja has pointed out, reduces the likelihood of at least one type of nightmare Singularity scenario.[334] As I argued in Chapter 3, the Singularity might go wrong by creating an unfriendly ultra-intelligent AI that annihilates us all to capture and preserve free energy. Such an AI would quickly make itself known by turning off or capturing the free energy from stars. So Fermi's paradox provides some evidence that the Singularity won't spawn a free-energy-gobbling monster.

2. AI is Too Hard

Not in a trillion years (absent further evolution) could monkeys bring about a Singularity. Perhaps humans also don't have what it takes. Eric Baum, a physicist who studies machine language, writes:

> The human mind only occurred [because] evolution trained on data for 4 billion years, involving (as I estimate) perhaps 10^{35} or more separate learning trials, and in a sense distilled all this learning into a compact expression in the DNA.[335]

If it really takes anywhere near that much effort and trial and error—10^{35} is a huge number—then humanity is far, far away from being able to create AI. But, I believe, we have talents, such as the ability to devise technologies to scan the human brain, that evolution lacks. Convergent evolution also provides some evidence in favor of our being able to produce AIs.

Imagine that there exists a special cave that over a long period of time a huge number of people have searched for. If only one person finds the cave, it was probably well hidden. But the more people who independently locate the cave, the less concealed we should expect it to have been. Now think of the cave as the genes necessary for "complex learning, memory, and tool use" and the evolutionary forces operating in different creatures as the searchers.

As Carl Shulman and Nick Bostrom write, across different species evolution independently produced "complex learning, memory, and tool use both within and without the line of human ancestry."[336] Understanding convergent evolution should reduce our estimate of how hard it must be to develop the part of humanity's intellectual toolkit shared by reasonably bright creatures such as octopuses, elephants, and crows.[337] If the blind forces of evolution could find this toolkit several times, mankind should be able to "find it" as well.

Still, this leaves open the possibility that the part of our intellectual toolkit that no other species possesses only came about because of some astronomically unlikely freak event that human researchers will likely never discover. If this is the case, however, neuroscientists will eventually stumble upon some roadblock, such as a type of brain cell that does something that defies analysis. Every day that passes without neuroscience identifying this kind of roadblock reduces the odds of our brains being too difficult for us to understand. Also, even if this roadblock exists, so long as programmers could create a computer simulation of it we should still achieve a Singularity.

Science keeps revealing more and more of how the universe functions. It's possible that someday this will stop, but the long-term trend favors us figuring out how most of the physical world, including the human brain, operates. This is especially true since the equipment we use to study the brain continually improves.

3. The Yudkowsky/Hanson Debate

Eliezer Yudkowsky believes that a Singularity will probably come about through an intelligence explosion, while Robin Hanson thinks it will likely materialize through emulations. Yudkowsky and Hanson have extensively debated their views. I present below a taste of their arguments. If both men are right in their criticisms of each other, then a Singularity is much less likely to occur.

Eliezer's primary objection to the feasibility of brain emulations is that similar technologies haven't been built by reverse-engineering biology—when humans learned to fly, we did it by building airplanes, not by trying to simulate birds. As he wrote, "Cars don't emulate horse biochemistry, sonar doesn't emulate bat biochemistry, compasses don't emulate pigeon biochemistry, suspension bridges don't emulate spider biochemistry, dams don't emulate beaver building techniques, and certainly none of these things emulate biology without understanding why the resulting product works."[338] A continuation of these trends would mean that we won't create AI by simulating the human brain, but rather by designing AI essentially from scratch.

As Hanson told me, the implausibility of some James Bond villains illuminates a reason to be skeptical of an intelligence explosion.[339] A few of these villains had their own private islands on which they created new, powerful weapons. But weapons development is a time- and resource-intensive task, making it extremely unlikely that the villain's small team of followers could out-innovate all the weapons developers in the rest of the world by producing spectacularly destructive instruments that no other military force possessed. Thinking that a few henchmen, even if led by an evil genius, would do a better job at weapons development than a major defense contractor is as silly as believing that the professor on *Gilligan's Island* really could have created his own coconut-based technology.

If the US military wanted to build a laser rifle, it wouldn't just give the assignment to the smartest MIT graduate. Rather, it would offer to pay billions of dollars to some defense contractor. This contractor would use the billions to hire thousands of employees and several defense subcontractors, each of whom would rely on the efforts of their own employees and the work of many other firms. It's exciting and romantic to imagine a single scientist working alone, inventing new products that change the world, but this isn't how innovation has proceeded. If John von Neumann had created the computer and the atomic bomb all by himself, then it would seem obvious that an AI even a little bit smarter than him could initiate an intelligence explosion, but even the mighty von Neumann achieved success only through collaborating with many others. Perhaps, just as it is absurd to imagine an isolated human being making major technological breakthroughs in a short period of time, it's unreasonable to predict that a single AI could undergo an intelligence explosion in part by making discoveries that had so far eluded the rest of the world.

Think of an innovation race between a single AI and the entirety of mankind. For an intelligence explosion to occur, the AI has to not only win the race, but finish before humanity completes its next stride. A sufficiently smart AI could certainly do this, but an AI only a bit brighter than von Neumann would have not the slightest chance of achieving this margin of victory.

Let's revisit the world of James Bond to locate another reason to be skeptical of an intelligence explosion. It's unlikely that a villain's new superweapon could be assembled from parts purchased at Walmart and Home Depot. Rather, the weapon would likely require custom-made components necessitating that the villain's island contain a large manufacturing base.

Analogously, as an AI got smarter, it would no doubt need to make its own components to get even smarter. Initially, the AI could likely increase its processing power by paying for already existing cloud computing. But eventually an AI racing to superintelligence would reach the limit of what existing products could give it and would need to manufacture its own computer equipment. One of economists' favorite stories, *I, Pencil*, written by Leonard E. Reed in 1958,[340] shows the challenge of this.

I, Pencil describes the vast number of people needed to make a single pencil. To manufacture a pencil you need to cut a tree, which requires you to have a saw, which requires you to have metal. The metal must be found in the ground, and then dug up with other tools that someone has to make. After the tree is cut down, it must be transported to a factory on vehicles, which themselves must be made from materials acquired by people with tools. All the workers in this process must be fed, and the food used to feed these workers must be grown with additional tools. *I, Pencil* shows the interdependencies of the economy.

You don't succeed in today's economy by just working by yourself; you need to get other people to change what they're doing to help you. The iPod could not have been built without Apple setting up a long supply chain. Each of these suppliers, in turn, required assistance from additional companies. Workers in these many companies had to get training and learn new skills to produce the iPod. This is the pattern that I believe has occurred with every significant technological development since the Industrial Revolution. An intelligence explosion would break with this pattern of interconnectivity, although a complete mastery of nanotechnology would allow an extremely intelligent AI to easily manufacture its own components.[341] But here we might run into the proverbial chicken-and-egg problem. To get a lot smarter, the AI might need to master nanotechnology, but to master nanotechnology the AI would have to be a lot smarter. Still, even understanding all of this, many in the Singularity community do believe in the likelihood of an intelligence explosion.

4. Cognitive Biases

Rationally predicting the future requires combating some destructive cognitive biases—the circumstances under which the human brain has a predisposition to wallow in irrationality. Consequently, the Singularity community's possible failure to defeat these biases represents another possible reason why you should doubt the conclusions of this book. The community recognizes the danger of cognitive biases, and through the blog Less Wrong, many members of it study how to overcome them. It's certainly possible, however, that the community hasn't overcome enough of its innate biases to make our predictions trustworthy.

For futurists, one of the most dangerous cognitive biases was uncovered by Daniel Kahneman, winner of a Nobel Prize in economics. Kahneman conducted an experiment in which he told his test subjects to imagine that:

> Linda is 31 years old, single, outspoken, and very bright. She majored in philosophy. As a student, she was deeply concerned with issues of discrimination and social justice and also participated in antinuclear demonstrations.[342]

The subjects were then asked to rank a set of statements by the likelihood of their being true. Two such statements were:

1. **Linda is a bank teller.**
2. **Linda is a bank teller and is active in the feminist movement.**

Many of the subjects ranked (2) ahead of (1), even though this is logically impossible. It must be more likely that Linda is a bank teller than that she is a bank teller and is also something else. Saying that (2) is more probable than (1) is analogous to claiming that there's a greater chance that Linda has a son than that she has a child.

Kahneman performed a similar experiment in 1982 on participants at an international conference on forecasting.[343] Most of the participants were professional analysts. He divided the subjects into two groups and asked the first to estimate the likelihood that there would be:

1. **A complete suspension of diplomatic relations between the United States and the Soviet Union sometime in 1983.**

He asked the second group to predict the probability of:

2. **A Russian invasion of Poland and a complete suspension of diplomatic relations between the United States and the Soviet Union, sometime in 1983.**

The second group gave a higher probability estimate than the first, even though the first statement has to be more likely than the second. Kahneman showed that humans have an innate bias to give credence to a scenario when more details are attached to the scenario. This bias creates a danger for futurists, because, if unchecked, it will cause us to construct elaborate, incorrect descriptions of what the future might bring.[344] My prediction that robots will become as smart as people might *seem* more plausible if I were to add to it a story about how these robots will cause massive unemployment, which would destabilize the democracies in Europe, causing a few Eastern European nations to abandon their market economies and return to communism, all the while pushing at least one Western European country to become fascist, and so on and so forth.

Freedom is dangerous for prognosticators because it allows them to tell stories that are plausible but wrong. One way a futurist could exploit this freedom is by turning to the art of rhetoric and embellishing his predictions with particulars to make them seem more believable than they should, taking advantage of his readers' bias in favor of detailed predictions.[345] I don't follow this path, admittedly, in part, because many in the Singularity community study cognitive biases to improve their ability to understand the world.[346]

The approach I mostly go with is to explain that there are many possible paths the future could take, and that I'm not predicting which one we will follow, but that on many of these paths we will encounter enhanced humans and artificial intelligences. What I'm doing is analogous to saying that Linda is probably one or more of the following: a feminist, a socialist, a community organizer, a homeless advocate, a radical environmentalist, or a gay-rights backer. And since none of these groups votes Republican, Linda almost certainly doesn't contribute to the Republican Party. Because the details of the different stories I tell contradict each other, I think I'm counteracting the bias identified by Kahneman. At the very least, informing you of this bias should reduce its hold over you.

Kurzweil overcomes the prognosticators' problem of excessive freedom by basing his predictions on heavily documented, quantitative trends. Kurzweil's book *The Singularity Is Near* (which has 106 pages of endnotes) carefully tracks trends that show exponential improvements in the quality and price of many information-based products, such as computer chips, computer memory, DNA sequencing, nanotechnology, brain scanning, and machines in the human body.[347]

Hanson, who writes the blog *Overcoming Bias*, suspects that at least two other biases prejudice the views of many members of the Singularity community. Hanson told me that we should be suspicious if a group of people with trait X argue that in the future trait X will be all-important because this group might be making predictions about the future to raise the status of people with trait X, or this group might have an irrationally high opinion of trait X. Lots of Singularitarians have extremely high measured intelligence.

Hanson thinks that some futurists also have a cognitive bias toward expecting "an unrealistic degree of [self-sufficiency] or independence."[348] For most of mankind's existence, we lived in small, autonomous hunter-gatherer tribes. Evolutionary selection pressures haven't had time to adjust our brains to the fact that we now live in an extraordinarily interconnected world in which no one small group of people can really do all that much to change the existing social order, especially over a short period of time. (Even the leaders of large organizations, such as Steve Jobs was, don't influence society alone; their achievements, too, depend on the efforts of many.) Hanson writes that failure to take this fact into account might be a reason many members of the Singularity community believe there is a significant chance that there will be an intelligence explosion shortly after an AI achieves human-level competence.

WE WANT TO FEEL SPECIAL

If the Singularity is near, then the next century will determine whether humanity dies out or spreads throughout the universe. In the wake of a successful Singularity, trillions upon trillions of people will be born, and they will look back on our time as the most important in human history, since it decided the fate of our species. But this would mean we are extremely special. In general, humans have a desire to feel special, so you should be distrustful of a group of people, such as Singularitarians, who have formulated supposedly logical reasons why they are special. Also, since you are probably not special, you should be distrustful of any theory that conjectures that you are indeed special.

Rapture of the Nerds

An Internet quote attributed (apparently falsely) to famed seventeenth-century mathematician Blaise Pascal (who did make a similar point in different words) states that "There is a God-shaped vacuum in the heart of every man which cannot be filled by any created thing, but only by God, the Creator, made known through Jesus." I think the quote is right in that the vast majority of humans want to believe in a supernatural world in which eternal life is possible. The views of many Singularity believers do have a strong religious flavor. After all, Singularitarians fear that an unfriendly ultra-AI (Devil) will destroy mankind (send us to hell) and think that Yudkowsky and the Singularity Institute he founded (church/cult) represents our best chance at survival (salvation), and that if we do survive we will have an excellent chance of living for a very long time with a friendly ultra-AI in utopia (spending eternity with God in heaven). The Singularity has been mocked as "the Rapture for nerds."[349]

Singularity believers claim that their views do not arise from faith but rather come from their reasoning about the future direction of technology. Critics, however, might counter by saying that lots of humans have trouble reasoning about the supernatural, which is why we have so many mutually contradicting religions, and consequently Singularity believers should not trust their ability to reason about the Singularity. Singularity believers' response should be that they have made tremendous efforts to understand what it means to be rational and to study the cognitive biases that inhibit rationality, and so you can have a reasonable level of trust that their beliefs about the Singularity are reason-centered.

Yudkowsky had a witty response when a dinner party guest proclaimed, "I don't believe artificial intelligence is possible because only God can make a soul." Yudkowsky retorted, "You mean if I can make an artificial intelligence, it proves your religion is false?"[350]

More likely than not, most folks who die today didn't have to die![351]
—Robin Hanson

CHAPTER 18
SINGULARITY WATCH

Do young programmers at Google not save for retirement because they expect a Singularity to arise before they turn sixty? Do more and more self-made tech multimillionaires contribute to organizations such as the Singularity Institute for Artificial Intelligence? It makes sense to watch where the smart people in the high technology community put their money for clues to the odds of a Singularity occurring.

Although an intelligence explosion could, literally, happen tomorrow, we will most likely get an AI-centered Singularity only after considerable improvements in computing hardware. If programmers never master massive parallel processing, then the hardware metric that most matters is the speed of the fastest computer. With parallel processing, however, an AI could be run on a huge number of relatively slow computers, provided they were cheap enough; so the hardware factor that should top your Singularity watch list is the quantity of computing power you can buy per dollar.

Computers get faster and cheaper only because of huge investments in research and development made by firms like Intel. Hardware manufacturers will make such expenditures only if they expect a high future demand for computing power. Consequently, consumers' desires for categories of products that rely on ever-increasing quantities of computational resources also deserve a prominent place on your Singularity watch list.

The Pac-Man video game of my youth looks pathetic compared to today's best-selling games—an advance made possible by Moore's Law. But how much room for improvement is there in video games? If, in another decade, games have almost all the computational resources they will ever need, then the Singularity will likely be further away than if games keep gobbling processing power-ups.

It seems near certain that virtual reality will someday sweep the marketplace. But, I wonder, how far away are we from having good-enough computing hardware to make virtual reality look like actual reality? Learning that the technology will, over the coming decades, consume ever-increasing amounts of computing power should convince you that the Singularity is a bit nearer than if the contrary holds.

Data analysis, I suspect, will never become saturated with computing power. The movements of atoms, people, and financial instruments are extremely complex, and I'd bet that long before we have computers powerful enough to fully simulate any of these movements, we would have machines with the computational capacity to give us a Singularity. Take it as a bit of a sign of an approaching Singularity if you read articles discussing how, with much, much faster computers, it would be possible to design vastly better drugs or materials.

Robots have the potential to be another huge source of demand for computing hardware. And, of course, robots would necessarily have at least limited artificial intelligence. I doubt much time would elapse between the creation of Rosie, the robot maid on the 1960s TV show *The Jetsons*, and a Singularity.

Similarly, I believe we would have a Singularity thrust upon us very quickly after someone creates an AI like HAL from the movie *2001: A Space Odyssey*. Any artificial general intelligence such as HAL could almost certainly become much smarter and more capable just by running on faster or more numerous computers. Consequently (and perhaps tragically), I strongly suspect that:

HAL + Continued Exponential Growth in Computing Power = Not-Too-Distant Singularity

Brain implants that can raise the general intelligence of a healthy person would be a strong sign that mankind is near a Singularity. If our brains can be improved by direct implants, then it seems likely that as computers become faster and smaller, we will keep getting brighter. Furthermore, the benefits of intelligence-enhancing brain implants would be so enormous that expected future demand for the implants would spur computer hardware companies to make huge investments in research and development.

With regard to human intelligence, learning that additive genes explain at least half of the variance in human intelligence would be a positive indication of the potential of genetic manipulation to create significantly smarter people than have ever existed. Another key indicator would be research showing that having many high-IQ genes usually doesn't create health problems, or at least that any resulting health challenges can be fixed by modern medicine.

Parents, however, might not advertise the steps they took to raise their kids' IQs, so look for children accomplishing astonishing things at extraordinarily young ages. If, for example, a ten-year-old Chinese child makes a significant contribution to physics, then suspect that eugenics will soon birth children with the potential to remake civilization.

Once you have accepted that a Singularity probably is near, you should decide how to personally prepare for it.

Personal Preparations

When choosing a career or contemplating starting a business, you should attempt to predict how future intelligence enhancements will impact different trades and professions. The good news is that, at least for a while, most enhancements will make the average person richer and so will increase the profitability of the majority of professions and industries.

Knowing about intelligence enhancements will make you a better stock market investor. Financial markets offer the highest rewards to those who act on information that the broad market hasn't yet taken into account. If the premise of this book is correct, you have inside information on the most radical change that will ever befall humanity. Profit from this knowledge!

Many people invest in stocks to have financial security when they retire. If you're saving for a retirement that won't start for a few decades, you should consider the value money will have post-Singularity and how the Singularity will influence the rate of return on investments. Your retirement, however, might not be as near as you think because enhancements such as brain-boosting drugs will affect how long you will be mentally capable of staying at your job.

Even workers who intend to never use cognitive-enhancing drugs should think through their effects. If, for example, you're a lawyer in a big law firm and can't figure out how a fellow employee manages to put in such long hours on such tremendously tedious tasks, the answer might be Adderall or modafinil. The performance difference between drug users and abstainers will almost certainly grow. After extremely effective memory-improving drugs come out, it might become impossible to rise to the top of professions that place huge demands on memory (e.g., tax law) without either using the drugs or being born with a photographic memory.

Parents need also to be concerned with brain-boosting drugs. Even if your kid never uses them, the child down the street who's in competition with your kid for class valedictorian might. If you put lots of pressure on your child to excel academically, he might turn to your school's black market in Adderall for help. If you had a frank talk about drugs with your kid that didn't mention Adderall, you haven't given him all of the advice and warnings he needs.

Parents should estimate how intelligence enhancements will affect the kinds of skills their children will eventually require. Will fluency in foreign languages matter if we have extremely proficient translation programs or if memory-improving pills soon make learning a foreign language "a trifle?" Parents should keep an eye on brain-fitness research to determine whether playing a few hours of certain kinds of video, games such as the dual *n*-back, can permanently increase children's working memory and IQ.

Singularity expectations should give hope to parents with severely disabled children. A good Singularity would almost certainly bring forth both the desire and the knowledge to cure your child's afflictions. Even if your offspring must suffer for fifty years with a horrible illness, if he survives to a positive Singularity, he will probably be able to live many times that many years more in a world in which he could choose to be smarter, healthier, and happier than any human being alive today.

If you have suicidal thoughts because life currently seems too painful to bear, take into account that if you make it to a positive Singularity, you will likely spend well over 99 percent of your life living in times of great happiness. A good Singularity will make everything and everyone a lot better.

Even my old or unhealthy readers who have less than a decade left of their "natural" life spans have a realistic shot of making it to the Singularity if they pay heed to what I now advise. In the introduction, I wrote, "This book has one recommendation that, if followed, could radically improve your life." I will now fulfill my introductory promise. My potentially life-changing advice: after your legal death, your body will be burned, buried, or frozen. Go with frozen.

AMBULANCE RIDE TO THE FUTURE

I have signed a contract for an arrangement with the cryonics provider Alcor. If I found out I would soon die, I would call Alcor and they would send a team out to me. After I died, this team would then immediately lower my body temperature by bathing my head in ice, then transport me to Alcor's Arizona facility. In case I suffer an accidental death, I wear a necklace instructing medical personnel to call Alcor if they cannot revive me.

Once Alcor has my body, the company will use liquid nitrogen to cool me to extremely low temperatures to prevent bodily degradation. I will, hopefully, remain cryogenically preserved until the technology exists to revive me. I fully believe that a friendly ultra-AI could revive me with trivial effort.

Cryonics is based on the understanding that there is a difference between legal death and actual death. You are not considered legally dead just because your heart stops working because hearts can sometimes be restarted, causing a "dead" person to come back to life. Since we don't have the technology to revive the brain, however, once your brain stops functioning completely you are considered legally dead. But cryonics enthusiasts believe that if your brain is intact enough for future nanotechnology to restore it to its pre-frozen state, then you shouldn't be considered dead merely because your brain temporarily ceases functioning.

Eliezer Yudkowsky once described cryonics as "an ambulance ride to the future."[352] If I am preserved in liquid nitrogen, sufficiently powerful nanotechnology could enter my body and repair every single damaged cell. We can be almost certain that such technology is possible, at least in theory, because natural human healing does essentially the same thing: when the body is damaged, our proteins repair the damage at the cellular level. Eric Drexler, often called the "father of nanotechnology," has written that the odds of cryonics revival technology being possible "seem excellent."[353]

Unfortunately, cryogenic cooling damages the brain. Consequently, the restorative nanotechnology that revives me would have to repair the harm caused by the preservation in addition to fixing whatever killed me.

As of this writing, medical science is close to being able to preserve kidneys at cold temperatures for transplant. The frozen hibernation of the frog *Rana sylvatica* is also evidence for the future practicality of cryonics. When this species of frog freezes,

> [its] heart stops completely. Ice encases the organ and forms inside the muscle and the chambers. There's no need for a working pump now because the blood, too, is frozen. This state of arrested heart function can be tolerated for many days and perhaps months, but upon thawing the contractions spontaneously resume.[354]

Admittedly, however, preserving a kidney or a frog is much easier then preserving a brain, due to complexity and size factors.

Cryonic revival would fulfill a wish of inventor Benjamin Franklin, who wrote:

> I have seen an instance of common flies ... drowned in Madeira wine ... [h]aving heard it remarked that drowned flies were capable of being reviv'd by the rays of the sun. ...
>
> I wish it were possible ... to invent a method of embalming drowned persons, in such a manner that they may be recalled to life at any period, however distant; for, having a very ardent desire to see and observe the state of America an hundred years hence, I should prefer to any ordinary death, the being immersed in a cask of Madeira wine ... till that time, to be then recalled to life by the solar warmth of my dear country![355]

Alcor charges members $800 a year while the member is legally alive and a one-time fee of $200,000 to have the whole body preserved, or $80,000 to have just the head preserved. Most cryonics members buy life insurance policies to cover the final fee. One healthy eighteen-year-old I know pays $21 each month for $250,000 in life insurance coverage.[356] The Cryonics Institute, the only other large American cryonics provider, charges considerably less than Alcor.

Many members also have separate life insurance for their family. In the past, some of the families of legally dead cryonics members have tried to stop the preservation process so they could keep funds that would otherwise go to the cryonics provider. The advantage of paying for cryonics with a life insurance policy is that the cryonics organization legally owns the policy, so family can't contest a will to get the insurance payout for themselves.

Besides the cost savings, the benefit of having merely your head frozen is that it's easier to preserve just the brain than the entire body, so the brain-only preservation process does less damage to neurons. I, however, intend to have my entire body preserved because I suspect that the non-brain part of me plays an important role in my personality by, for example, regulating brain-influencing hormones.

If an ultra-intelligent AI could travel back in time to bring back the dead, then cryonics would gain me nothing. But since time travel might be prohibited by the laws of physics, whereas strong nanotechnology almost certainly isn't, cryonics does significantly increase my chances of being alive a thousand years from now.

At least six people mentioned in this book have signed up with either Alcor or the Cryonics Institute:

- Ray Kurzweil, inventor and leading Singularity intellectual[357]
- Peter Thiel, self-made technology billionaire[358]

- Robin Hanson, economist
- Eliezer Yudkowsky, leading friendly-AI theorist
- Michael Anissimov, Media Director for the Singularity Institute
- Eric Drexler, the father of nanotechnology[359]

Furthermore,

- super-genius Thomas McCabe, who helped edit this book, told me he was in the process of completing the paperwork at the time I talked with him;
- Skype cofounder Jaan Tallinn said that although he isn't currently signed up, he would use cryonics if he or a loved one were about to die; and
- former Alcor president Stephen Van Sickle told me that there are one or more famous businesspeople, not associated with the Singularity movement, who are [secret] Alcor members.

Singularitarians are the only group that cryonicists have really won over. Probably in excess of 100 million people have heard of cryonics, yet the cryonics movement has failed to attract many adherents. As of this writing, Alcor has only 968 members and 111 preserved patients. The Cryonics Institute has a few more than 1,000 members and 111 preserved patients. I'm frustrated that out of the millions of people who have surely heard of cryonics, fewer than two thousand have joined a cryonics organization.

I could learn a lot about your beliefs and preferences by asking you questions. Based on your answers, I could predict what sorts of behavior will make you the happiest. For example, if you answered "yes" to the questions "Do you enjoy thrills?" and "Do you like going very fast?" I would be justified in recommending that you go on a roller coaster ride.

I'm now going to use this Socratic questioning method on you. I have encountered many of the same objections to cryonics over and over in the numerous conversations I have had on the subject, and I'm going to assume that you object to cryonics for one or more reasons I have heard before. Below this paragraph is a list of cryonics objections, and beneath each objection is a question. For reasons that either I will provide or that should seem obvious, giving answer (B) to any question means that its corresponding objection shouldn't block you from joining the cryonics movement.

Objection 1: Cryonics is unnatural.

Question 1: Would you support a law prohibiting all medicine not used by our hunter-gatherer ancestors?

A) Yes

B) No

Objection 2: Once you have died, you are dead.

Question 2: You fall into a lake while ice skating, and your body quickly freezes. A year from now your body thaws out, and for some crazy reason, you think, look, and act just as before. Are you alive?

A) No

B) Yes

Objection 3: Even if cryonics works, the "person" that would be revived wouldn't really be me.

Question 3: Were you yourself ten years ago?

A) No

B) Yes

If you answered "Yes" to Question (3), then you are not defining "you" by the physical makeup of your body because almost none of the atoms in your body today are the same as they were ten years ago. Your body has undergone many changes over the last decade, meaning that if you still identify yourself as you, you must be defining yourself by some broad structure and not merely by the exact arrangement of the molecules that compose "you."

Also, imagine that a year after joining Alcor, you wake up one morning in a hospital bed and see the smiling face of your grown-up child. Although thirty years have passed since you died in your sleep, no subjective time has transpired, and your body has the exact same look and feel as it did before you went to bed. Indeed, had Alcor placed you back in the room in which you died, you would think today was a normal morning. Are you still you? Are you glad you signed up with Alcor?

Furthermore, consider two forty-year-olds named Tom and Jane. Tom legally dies in 2020, is cryogenically preserved, and is then revived in 2045 by a process that restores his body and brain to the condition it was in before he legally died. Jane survives to 2045. The Tom of 2045 is vastly more similar to the Tom of 2020 than the Jane of 2045 is to the Jane of 2020, as the Jane of 2045 has undergone twenty-five extra years of aging and life experience. So if you believe that Jane has stayed Jane over that time period, you should certainly think that the post-cryonics Tom is the same as the pre-cryonics one.

Objection 4: I don't want to wake up a stranger in a strange world.

Question 4: While driving, you get into an accident. When you wake up in a hospital, an FBI agent tells you that the son of a Mafia leader died in the accident, and although the accident wasn't your fault, the leader will hold you responsible. If the Mafia thinks you survived the accident, then they will kill you. The agent confesses that the Mafia has infiltrated the FBI, and so the government will never be able to protect you. The agent provides one option for survival. He will fake your death and make people think your body was burned beyond recognition. The agent will then give you a new identity and transport you to another country, one very different from your own. You will never be able to contact any of your old friends or family again because the always suspicious Mafia will monitor them. Do you accept the agent's offer?

A) No

B) Yes

Plus, if you are revived in a world you would rather not live in, you will probably have the option of killing yourself.

Objection 5: If revived in the future, I wouldn't have any useful skills.

Question 5: You have a fatal disease that has only one cure, but this cure costs $1 billion, which you can't possibly raise. NASA, however, makes you an offer. The space agency is launching a rocket that will travel near the speed of light. Because of Einstein's theory of relativity, although the mission will take you only one subjective year to complete, when the rocket returns to Earth, one thousand years will have passed. Because you are the most qualified person to fly the rocket, NASA will pay the cost of your disease's cure if you accept the mission. Do you accept?

A) No

B) Yes

Also, you are only likely to get revived in a friendly rich world in which the least fortunate are still much better off than most people are today. If a utopian Singularity has occurred, then everyone alive in that time will be better off than anyone is today. Furthermore, the nanotechnology that revives you would almost certainly be able to make you strong and healthy enough to work. Finally, this nanotech would have a high chance of allowing you to alter your brain so that you could quickly learn new skills.

Objection 6: The people who revive me might torture me.

Question 6: If you knew an intelligence explosion would occur tomorrow, would you commit suicide today to avoid the chance of being tortured?

A) Yes

B) No

Objection 7: I don't cherish my life enough to want to extend it with cryonics.

Question 7: You will die of cancer unless you undergo a painless medical procedure. Do you get the procedure?

A) No

B) Yes

Objection 8: It would be morally superior for me to donate money to charity rather than spending money on cryonics.

Question 8: Same as Question (7), but now the operation is expensive, although you can afford to pay for it.

Also, most readers of this book could pay for cryonics by spending less on personal consumption, thereby not reducing the amount they give to charity.

Objection 9: It's selfish of me to have more than my fair share of life, especially since the world is overpopulated.

Question 9: Same as Question (7), except that your age is well above the number of years the average human lives.

Objection 10: I believe in God, the real one with a capital G, not an extremely smart artificial intelligence. I don't want to post-pone joining him in the afterlife.

Question 10: Same as Question (7).

Also, even the extra million years of life that cryonics might give you is nothing compared to the eternity you believe you will eventually spend in heaven. If you believe that God wants you to spend time in the physical universe before joining him, might he not approve of you using science and reason to extend your life, so you can better serve him in our material world?

Many faiths believe it's virtuous to have children, raise these children to understand God, and then hope these children beget more children who will carry on the faith. If there is a God, he appears to have started us out on a tiny planet in an empty universe. People who make it to the Singularity would likely get to "be fruitful and multiply" and populate God's universe. (Don't worry if you are past childbearing age: any technology that could revive the cryogenically preserved could be used to help you have children).

Objection 11: If people find out I have signed up for cryonics, they will think I'm crazy.

Question 11: Same as Question (7), but now most people think the type of operation you will get is crazy.

The stigma of cryonics is real and has even made this author nervous about outing himself for fear that it might make it harder for me to find alternative employment if I should choose to leave or get fired from Smith College. You could always keep your membership secret.

Objection 12: Cryonics might not work.

Question 12: Same as Question (7), but now the procedure only has a 5 percent chance of saving your life.

At the end of the book, I'll ask you a final question.

CRYCRASTINATION[360]

Far more people intend to sign up for cryonics than have actually done so. Many figure that being young and healthy means they have lots of time to wait. But a sudden, unexpected death terminates most hope for cryonics.

Furthermore, a procrastinator struck by a fatal disease might find it financially impossible to pay for cryonics. Since life insurance companies won't sell an inexpensive policy to someone with a terminal disease, a man who put off signing up for cryonics until after he was gravely ill would have to pay for cryonics out of pocket. But such a man might have a greatly reduced income, combined with a need to pay for health care and to provide his family with sufficient wealth to cushion the financial loss of his death.

If you come from a wealthy family, don't count on them paying for your future preservation, since most people consider cryonics to be a waste of money. Even if your family promises to have you preserved, once you die they might renege on the promise, since they would likely see no benefit in fulfilling it.

BETTER LEADERS

Communist dictator Joseph Stalin maintained power by killing millions of his countrymen and terrorizing the rest. He often lashed out at his old comrades, sometimes killing them and their families; at other times, he was satisfied with just jailing their wives. Stalin, who was denounced shortly after his death by his successor Nikita Khrushchev, must have known how hated he was. But the dictator knew that those who hated him were too weak or fearful to hurt him.

The first cryonics patient was preserved in 1967, fourteen years after Stalin's death. But what if, I wonder, cryonics existed during the time of Stalin, and the dictator hoped to have himself preserved? Stalin was too smart to think his successors would have ever wanted him back. So if, at the beginning of his rule, Stalin had hoped to someday use cryonics, he would have had to be a less ruthless ruler. For any hope of cryonic revival, the world will need to want you back. So if the world's leaders intend to use cryonics, they will have to care more about what the future will think of them.

Widespread cryonics use could also improve governance by turning senior citizens into more future-oriented voters. In rich countries such as the United States, the government transfers huge amounts of money to the elderly. Governments acquire this money by taxing workers. These taxes, unfortunately, slow economic growth and probably postpone the Singularity. Today, a self-interested senior citizen would oppose the government cutting benefits to the elderly, even if the reductions would increase economic growth. But a senior hoping that cryonics will give him a second life might be more willing to accept short-term personal pain in return for a higher living standard in the far future.

Not only taxes but also medical costs would be lower if cryonics becomes a real option. Given the promise of living indefinitely in full health, governments and patients would be much less willing to pay for extremely expensive and painful medical treatments that extend life by only a few weeks.[361]

EVIL

My thinking on cryonics has progressed like this:

> Cryonics is crazy→Cryonics might work→I will sign up someday→I have signed up but am embarrassed by my choice→I have signed up and am proud of my decision→I hope my friends and

family sign up→Most people who reject cryonics do so for irrational reasons→Underuse of cryonics is the greatest evil of our time.

About 107 people die every minute, all too many of them children.[362] Routine
cryonics could give us a reasonable chance at saving most of them. Free cryonics, if done on a massive scale, could be offered to everyone, and the total cost
would likely represent a tiny fraction of the amount currently spent on health
care. There would be considerable medical cost savings if people who have brain-
wasting diseases such as Alzheimer's were frozen before they would normally die.

How will the future judge us if cryonics works and future historians come
to believe that we should have known that it would work? Will our descendants
think us monsters for letting so many die unnecessarily?

ADMITTEDLY, IT MIGHT NOT WORK

I'm far from certain that if I were to die today, I would someday be cryogenically revived. I might die in a manner that prevents Alcor from preserving my
brain. Even if Alcor gets my body, they might not successfully preserve me, due
to error or financial collapse. If the technology to revive me didn't arrive for at
least a thousand years, I'd give myself little chance of survival. But I think Alcor
is up to the job of keeping me alive until we reach the Singularity. Of course, a
post-Singularity civilization might not want to revive me—although I believe
a friendly AI would probably do it, and I certainly hope that after a positive
Singularity one of my living relatives would make strenuous thawing efforts on
my behalf to let me experience a utopian Singularity!

My last question for you:

> One minute from now a man pushes you to the ground, pulls out a
> long sword, presses the sword's tip to your throat, and says he will
> kill you. You have one small chance at survival: grab the sword's
> sharp blade, throw it away, and then run. But even with your best
> efforts, you will probably die. Do you fight?

If you are the kind of person who would fight and you find this book's
arguments about a Singularity plausible, then based on your own beliefs and
preferences, you should sign up for cryonics. If you do sign up because of me, I
would be grateful to receive an e-mail explaining which of my arguments you
found convincing.

ACKNOWLEDGMENTS

A big thank you to

Robin Hanson, Ray Kurzweil, Vernor Vinge, and Eliezer Yudkowsky—whose ideas and scholarship form the base on which this book stands.

Debbie Felton—for spending so much time helping her husband craft a book concerning a subject she didn't find inherently interesting.

Thomas McCabe—for providing a huge number of content and style suggestions.

Bruce Goldman—for helping me formulate some of the basic structure of this book.

Sam Fleishman—my agent, who worked tirelessly to get me the best contract with a supportive publisher.

I'm also grateful to the following people who spent time communicating with me about the content of this book.

Michael Anissimov	Jason Freidenfelds	James Middlebrook
Mike Baker	Lixin Gao	Alexander Felton Miller
Michael Bush	Katja Grace	Albert Mosley
Judith Cardell	Hank Greely	Steve Omohundro
Gregory Clark	Andrea Hairston	Eric Polizzi
Gregory Cochran	Joe Hardy	Jennifer Rodriguez-Mueller
Keith Curtis	Steve Hsu	Elizabeth Savoca
Amnon Eden	Earl Hunt	Carl Shulman
Glenn Ellis	Garett Jones	Jaan Tallinn
Tyler Emerson	Greg Laden	Peter Thiel
Rick Fantasia	Shane Legg	Dominique Thiebaut
Gary Felder	Moshe Looks	Michael Vassar

Finally, I'm grateful to the many commenters on the blogs Overcoming Bias and Less Wrong for making numerous insightful comments about artificial intelligence.

NOTES

1. Vinge (1993).
2. Miller (2007).
3. Frank (2011).
4. Vance (2010).
5. Walsh (2001); Hawking (2000).
6. Kurzweil (2005).
7. Kurzweil (2005).
8. Kurzweil (2005).
9. Legg (2010).
10. Elango et al. (2006). Sentence structure similar to sentence in Anissimov (2007).
11. Macrae (1992).
12. Macrae (1992).
13. Macrae (1992).
14. Macrae (1992), 29. Quoting Lewis Strauss.
15. Macrae (1992).
16. Macrae (1992).
17. Macrae (1992).
18. Macrae (1992).
19. Macrae (1992).
20. Macrae (1992).
21. Macrae (1992).
22. Hargittai (2006).
23. I have altered the photograph. http://www.lanl.gov/history/wartime/images/ProjectYBadges/v/vonneumann-john_r.gif
24. Ulam (1958).
25. Vinge (1993).
26. Singularity Institute (2011).
27. Vinge (2010).
28. Rumsfeld (2002).
29. Ray Kurzweil (2005).
30. Legend of the Ambalappuzha Paal Paayasam. This telling of the story is in my words.
31. Kurzweil (2005).
32. Kurzweil (2005).
33. Kurzweil (2005).
34. Kurzweil (2005).
35. Kurzweil (2005).

36. Kurzweil (2005).
37. Kurzweil (2005). Footnoted omitted.
38. Kurzweil (2005).
39. Kurzweil (2005). Footnote omitted.
40. Kurzweil (2005).
41. Kurzweil (2005).
42. Kurzweil (2005).
43. Kurzweil (2005).
44. Kurzweil (2005).
45. Kurzweil (2005).
46. Kurzweil (2005).
47. Polizzi (2010).
48. Lloyd (2006).
49. See Kurzweil (2005), 124 for an estimate of the speed of the human brain.
50. Kurzweil (2005).
51. Anissimov (September 15, 2011).
52. Good (1965).
53. Kurzweil (2005).
54. Kurzweil (2005).
55. Kurzweil (2005).
56. Kurzweil (2005).
57. Kurzweil (2005).
58. Kurzweil (2005).
59. Kurzweil (2005).
60. Hanson (January 31, 2011).
61. Hanson (June 20, 2011).
62. Kurzweil (2005).
63. Kurzweil (2005).
64. As Hanson (June 20, 2011) told me, you might also need to know about state changes in the cells, although this could take the form of an abstract model of states and not require a detailed understanding of all the cell's biochemistry.
65. http://www.alleninstitute.org/Media/documents/press_releases/2012_0321_PressRelease_ExpansionAnnouncement.html
66. Hanson (January 31, 2011).
67. Hanson (March 5, 2011).
68. Hanson (January 31, 2011).
69. Kurzweil (2005).
70. Legg (2010).
71. Yudkowsky (December 1, 2008).
72. Yudkowsky (December 2, 2008).
73. Yudkowsky (December 8, 2008).
74. http://singinst.org/summit2007/quotes/billgates
75. Zhou, Khemmarat, and Gao (2010).
76. Bostrom and Ćirković (2008).
77. Shakespeare, *Macbeth*, 7.1.46–47. Interpretation from Greely (2008b).
78. Kurzweil (2005).
79. Cochran and Harpending (2009).
80. Cochran and Harpending (2009).
81. Many of the ideas for this chapter came from Omohundro (2007 and 2010).
82. Tacitus (1999).
83. Zillmer et al. (1995).

84. Omohundro (2007).
85. Omohundro (2007).
86. Omohundro (2007).
87. Idea suggested to me by Tom McCabe, who edited most of this book.
88. Anissimov (January 15, 2011).
89. Berkshire Hathaway Inc. (2001).
90. Yudkowsky (December 22, 2007).
91. http://singinst.org/donors
92. Jaan Tallinn told me he was worth around $30 million. Tallinn (2010).
93. Yudkowsky (November 25, 2005).
94. Yudkowsky (September 15, 2008).
95. Yudkowsky (September 23, 2008) and Yudkowsky (September 24, 2008).
96. Yudkowsky (September 25, 2008).
97. Muehlhauser (2012).
98. Yudkowsky (2009). In some situations you do worse if people think you are rational than if people think you're irrational. But holding constant what other people think of you, you're always better off being rational than irrational.
99. Omohundro (2007).
100. Yudkowsky (September 3, 2010).
101. Yudkowsky (November 7, 2005).
102. Yudkowsky (September 3, 2010).
103. This creates a moral dilemma as to whether the AI should give weight to the welfare of the virtual Carl.
104. Adams (2011).
105. Salamon (2009).
106. Yudkowsky in Bostrom and Ćirković (2008).
107. Bostrom (2003b).
108. Nelson (2007) proposed something similar. I developed my scheme before learning of his.
109. Bostrom (2003a).
110. Hanson (June 26, 2001).
111. Yudkowsky in Bostrom and Ćirković (2008).
112. http://singinst.org/summit2007/quotes/jamaiscascio/
113. Yudkowsky (2001). Emphasis removed.
114. http://singinst.org/summit/; Hanson (March 14, 2008).
115. Homer (2000).
116. Gottfredson (2002).
117. Jones (2011b).
118. Gottfredson (2002).
119. http://mason.gmu.edu/~gjonesb/JonesADBSlides
120. Gottfredson (2005). Footnote omitted.
121. Lubinski et al. (2001).
122. Lubinski et al. (2001).
123. Lubinski et al. (2001).
124. Stanovich (2009). Footnote omitted.
125. Hunt (2011).
126. Gottfredson and Deary (2004).
127. Sabbah and Sheiham (2010).
128. Gottfredson and Deary (2004).
129. Arden et al. (2009).
130. Kanazawa (2011).

131. Engber (2012). Links omitted.
132. Gottfredson (2002).
133. Neisser et al. (1996).
134. Hunt (2011).
135. Hunt (2011).
136. Jones (2011b).
137. Intelligence Expert X (2010).
138. Hunt (2011).
139. Wai and Putallaz (2011).
140. Intelligence Expert X (2010).
141. Hunt (2011).
142. Heckman, Stixrud, and Urzua (2006).
143. Hunt (2011).
144. Deary et al. (2004).
145. Karlgaard (2005).
146. Gottfredson (1997).
147. Heckman et al. (2006). The article points out that noncognitive abilities do this as well.
148. Heckman et al. (2006).
149. I'm assuming an IQ standard deviation of 15 for this and all other IQ calculations used in this book.
150. Gensowski, Heckman, and Savelyev (2011).
151. Sanandaji (2001).
152. Jones (2011b).
153. Lynn and Meisenberg (2010), which Jones (2011b) cites.
154. Ram (2007).
155. Kanazawa (2006).
156. Sturgis, Read, and Allum (2010).
157. Jones (2011a).
158. Beaver and Wright (2011).
159. Caplan and Miller (2010).
160. Warner and Pleeter (2001).
161. Gates (2011).
162. Hunt (2011) and Hsu (2010).
163. Mutations can cause identical twins to have slightly different DNA.
164. Segal (1999).
165. Intelligence Expert X (2010).
166. Hunt (2011).
167. Caplan (2011).
168. Hunt (2010).
169. Davies et al. (2011).
170. Murray (2011).
171. Hsu (2010).
172. Hsu (2010) and Intelligence Expert X (2010).
173. Obama's mother had a PhD in anthropology and his father did graduate work at Harvard University.
174. Intelligence Expert X (2010).
175. Highfield (2007).
176. Cochran and Harpending (2009).
177. Cochran and Harpending (2009).
178. Cochran and Harpending (2009).

179. Cochran and Harpending (2009).
180. Cochran and Harpending (2009).
181. Buchanan (2011).
182. Frost and D'Anglure (2010).
183. Clark (2007).
184. By interest rates I mean real interest rates.
185. Clark (2007).
186. Clark (2007).
187. Clark (2007) wrote: "This is not in any sense to say that people in settled agrarian econo-mies on the eve of the Industrial Revolution had become 'smarter' than their counterparts in hunter-gatherer societies" (187), that would *seem* to contradict part of my thesis. But I subsequently wrote an e-mail to Clark which read in part: I would be grateful if I could say that I asked you if the following statement is true and you agreed that it was:

> Selection pressures in England from 1250-1800 probably increased the types of intel-ligences that make workers more productive in industrial enterprises. But this might have been due entirely to cultural transmission. Also, we can't conclude that the average citizen of 1800 England was smarter than his hunter-gatherer ancestor because there might be trade-offs between different types of intelligences with selection pressures increasing the types needed to be a successful farmer but decreasing the types needed to be a successful hunter-gatherer. Furthermore, hunter-gatherers perform tasks that intuitively seem beyond what most people in rich nations could efficiently accomplish.

Clark's response read in part:

> Your quote below exactly expresses what I would say. The long period of settled agrarian societies, with constant selective pressure towards people who succeeded economically, likely emphasized some human intellectual abilities, but diminished others. But the abilities that were enhanced would be the ones that would result in higher measured IQ.

The e-mails were sent on October 12, 2010 and October 13, 2010, respectively.
188. Clark (2007).
189. Clark (2007).
190. Yudkowsky (June 23, 2008). Hyperlink omitted.
191. This chapter greatly benefited from discussions with Carl Shulman.
192. BBC News (2009).
193. Vines (1997).
194. The die is numbered 1 to 100.
195. Let F(x) = the probability of getting an x or higher. Let x^* be the value of x such that if you roll the die twice the probability of getting x or higher on both rolls is 1%. $F(x^*)^2=1\%$ so $F(x^*)=10\%$. The probability of getting x^* or above on at least one roll if you have 24 rolls is $1-.9^{24}$. The probability that embryo selection as I've defined it will yield a genius is therefore $.1(1-.9^{24}) = 9.2\%$.
196. Intelligence Expert X (2010).
197. Levine (2010).
198. Levine (2010).
199. Levine (2010).
200. Miller (2000).
201. Nieli (2010), summarizing Espenshade and Radford (2009).
202. Hunt (2011).
203. Section informed by discussion with Carl Shulman. The breast example, however, is mine.
204. http://en.wikipedia.org/wiki/Cleft_palates as it appeared on June 6, 2011.
205. Hunt (2011).

206. BBC (2004).
207. Armstrong (2010).
208. Hazlett et al. (2005).
209. Wallis (2011).
210. Any similarities between this book's description of hyperlexia and the hyperlexia article in Wikipedia comes from this author having substantially edited the Wikipedia entry.
211. I draw this conclusion based on my extensive understanding of autism and my research on von Neumann. Von Neumann's well-above-average social skills and the pleasure he took from socializing make this an easy determination.
212. Moses (2009).
213. *Born on the Wrong Planet* (Hammerschmidt 2008) is the title of one book on autism; *Right Address . . . Wrong Planet: Children with Asperger Syndrome Becoming Adults* (Barnhill 2002) is the title of another. http://www.wrongplanet.net is the address of a "web community designed for individuals (and parents/professionals of those) with Autism, Asperger's Syndrome, ADHD, PDDs, and other neurological differences."
214. Lee (2000).
215. kjmtchl (2010).
216. kjmtchl (2010).
217. kjmtchl (2010).
218. http://www.sociopathworld.com/2009/08/sociopath-voices.html
219. Smith (2007).
220. Dikötter (1998).
221. I'm associating four top 1 percents with the first bullet point.
222. Frank (2011).
223. Cochran and Harpending (2009). Footnotes omitted.
224. Cochran and Harpending (2009).
225. Cochran and Harpending (2009).
226. Cochran and Harpending (2009).
227. Cochran, Hardy, and Harpending (2006).
228. Cochran, Hardy, and Harpending (2006).
229. Cochran, Hardy, and Harpending (2006).
230. Cochran, Hardy, and Harpending (2006).
231. Cochran, Hardy, and Harpending (2006). Cochran speculates that Gaucher might "increased axonal growth and branching."
232. Cochran (2010).
233. www.theuncertainfuture.com
234. Cochran (2010).
235. Cochran (2010).
236. Kaplan and Gellene (2007).
237. Moore (2011). Although amphetamine and methylphenidate are slightly different, I'm going to abuse terminology and use the term *amphetamine* to refer to both classes of drugs.
238. DeSantis, Webb, and Noar (2008).
239. Moore (2011).
240. Moore (2011).
241. Moore (2011).
242. Moore(2011).
243. British Medical Association (2007).
244. Moore (2011).
245. Caldwell et al. (1999).
246. Thirsk et al. (2009).

247. Müller et al. (2004).
248. Turner et al. (2003).
249. Kaplan and Gellene (2007).
250. Stough et al. (2011).
251. Sahakian and Morein-Zamir (2007).
252. Kling (2011).
253. Hardy (1992).
254. Armstrong (2010), quoting Thom Hartmann.
255. Moore (2011).
256. Kaplan and Gellene (2007).
257. Alzheimer's Association (2009).
258. Greely (2010).
259. Moore (2011). Quoting Nicolas Rasmussen.
260. Emonson and Vanderbeek (1995).
261. Greely et al. (2008).
262. Greely (2005–6).
263. Greely (2008a).
264. Greely (2005–6).
265. Not all double-blind drug trials give placebos to exactly half of their test subjects.
266. Moore (2011).
267. Hafner (2008).
268. Baker (2010).
269. Ignoring the time costs.
270. Jaeggi et al. (2008).
271. Hunt (2010).
272. The rules are a bit more complicated than this.
273. Jaeggi et al. (2008). See also Jaeggi et al. (2010). Moody (2009) critiques Jaeggi et al. (2008).
274. Wikipedia entry for fluid and crystallized intelligence as it appeared on December 22, 2010.
275. Jaeggi et al. (2008). But see Jaeggi et al. (2011), which found slightly different and harder-to-interpret results.
276. Hertzog et al. (2008).
277. Hardy, J. (2010).
278. Hertzog et al. (2008).
279. Fitzpatrick and Pagani (2012).
280. Morrison and Chein (2011).
281. Hardy, Q. (2010).
282. Robin Hanson suggested this to me.
283. Fromkin (2005).
284. Krugman (1997).
285. Martini (2001).
286. In 2004 I was the Republican nominee for the Massachusetts State Senate for Hampshire and Franklin counties.
287. *Buck v. Bell*, 274 US 200 (1927).
288. http://singinst.org/summit2007/quotes/rodneybrooks/
289. https://attra.ncat.org/intern_handbook/history.html
290. Cowen (2011).
291. Cowen (2011) and http://en.wikipedia.org/wiki/Advanced_Chess as it appeared on March 11, 2011.
292. I'm assuming that transaction costs are not so large as to eat up all of the gains from this trade.

293. Hanson (November 30, 2009).
294. Aaronson (2008).
295. This includes indirect agricultural production in which a country produces food by making non-edible items and trading them to other countries for food.
296. Ricardo (1821).
297. Idea from Robin Hanson.
298. Levy (2011).
299. Romney (2010).
300. Idea suggested to me by Robin Hanson.
301. Yudkowsky (November 22, 2008). Emphasis changed.
302. Shulman (2008).
303. Idea suggested to me by Robin Hanson.
304. Hanson (November 24, 2008).
305. Hanson (May 27, 2011).
305. Lat (2007).
306. Greely (2008b).
307. British Medical Association (2007).
309. Kurzweil (2005).
310. Clark (2007).
311. Boudreaux (2008).
312. Goetz (2010).
313. http://www.parkinson.org/Parkinson-s-Disease/Treatment/Experimental-Therapy--Clinical-Trials/23andMe
314. This estimate doesn't take into account cryonics.
315. http://www.fightaging.org/archives/2007/08/robert-bradbury-on-longevity-research.php
316. Median expenditures. http://nces.ed.gov/edfin/graph_topic.asp?INDEX=1. Data for 2007-2008. The exact median was $9,786.
317. Coulson (2008).
318. Dillon (2010).
319. Michael Anissimov suggested this example.
320. http://longbets.org/1/
321. Kirkwood (2010).
322. By energy I mean free energy.
323. Eichengreen and Mathieson (1999).
324. I've chosen not to reveal the manager's name out of concern for his privacy.
325. Adams (2007).
326. Gould (2004).
327. *Half Sigma* (2007).
328. Zhu, Lu, and Hesketh (2009).
329. Robinson and Wilson (2011.)
330. Grace (2010a). Katja Grace is the same person as Caitlin Grace.
331. Hanson (September 15, 1998).
332. Hanson (November 28, 2010, and April 1, 2011).
333. Hanson (March 22, 2010).
334. Grace (2010b).
335. Baum (2004).
336. Shulman and Bostrom (2012).
337. Shulman and Bostrom (2012).
338. Yudkowsky (November 10, 2007). Emphasis removed.
339. Hanson (June 20, 2011).

340. http://www.econlib.org/library/Essays/rdPncl1.html
341. Hanson (September 2001).
342. Tversky and Kahneman (1982).
343. Tversky and Kahneman (1983). *See* Yudkowsky (September 19, 2007).
344. Yudkowsky (September 19, 2007).
345. http://wiki.lesswrong.com/wiki/Dark_arts
346. www.overcomingbias.com and www.lesswrong.com
347. See Kurzweil (2005) for example.
348. Hanson (September 2001).
349. Derbyshire (2008).
350. Yudkowsky (July 31, 2007).
351. Hanson (December 12, 2008). Emphasis removed.
352. Yudkowsky (February 15, 2010).
353. Drexler (1986).
354. http://www.pbs.org/wgbh/nova/nature/costanzo-cryobiology.html
355. Franklin (1773).
356. Tom McCabe, who edited most of this book.
357. Philipkoski (2002).
358. He told me of his membership.
359. Pascal (2005).
360. Term told to me by Carl Shulman.
361. Yudkowsky (May 26, 2011).
362. https://www.cia.gov/library/publications/the-world-factbook/geos/xx.html. Estimate for July 2011.
363. Same person as Grace, Katja.

REFERENCES

Aaronson, Scott. May 13, 2008. "The Bullet-Swallowers." *The Blog of Scott Aaronson.* http://www.scottaaronson.com/blog/?p=326.

Adams, Scott. 2007. *Stick to Drawing Comics, Monkey Brain! Ignores Helpful Advice.* New York: Penguin Group.

Adams, Scott. January 27, 2011. "Comparing." *The Scott Adams Blog.* http://dilbert.com/blog/entry/comparing/.

Alzheimer's Association. 2009. "2009 Alzheimer's Disease Facts and Figures." http://www.alz.org/national/documents/report_alzfactsfigures2009.pdf.

Anissimov, Michael. June 4, 2007. "Response to Cory Doctorow on the Singularity." *Accelerating Future* (blog). http://www.acceleratingfuture.com/michael/blog/2007/06/response-to-cory-doctorow-on-the-singularity/.

Anissimov, Michael. January 15, 2011. "Yes, the Singularity is the Biggest Threat to Humanity." *Accelerating Future* (blog). http://www.acceleratingfuture.com/michael/blog/2011/01/yes-the-singularity-is-the-biggest-threat-to-humanity/.

Anissimov, Michael. September 15, 2011. "Interview with New Singularity Institute Research Fellow Luke Muehlhauser." *SIAI Blog.* http://singinst.org/blog/2011/09/15/interview-with-new-singularity-institute-research-fellow-luke-muehlhuaser-september-2011/.

Arden, Rosalind, Linda S. Gottfredson, Geoffrey Miller, and Arand Pierce. 2009. "Intelligence and Semen Quality are Positively Correlated." *Intelligence* 37 (3): 277–82.

Armstrong, Thomas. 2010. *Neurodiversity: Discovering the Extraordinary Gifts of Autism, ADHD, Dyslexia, and Other Brain Differences.* Philadelphia: Da Capo Press.

Baker, Mike. Interview with James D. Miller, October 6, 2010.

Barnhill, Gena P. 2002. *Right Address . . . Wrong Planet: Children with Asperger Syndrome Becoming Adults.* Shawnee Mission, KS: Autism Asperger Publishing Company.

Baum, Eric B. 2004. *What is Thought?* Cambridge: MIT Press.

BBC News. 2004. "Brilliant minds linked to autism." January 8. http://news.bbc.co.uk/2/hi/health/3380569.stm.

BBC News. January 9, 2009. "Breast Cancer Gene-Free Baby Born." http://news.bbc.co.uk/2/hi/health/7819651.stm.

Beaver, Kevin M., and John Paul Wright. 2011. "The Association between County-Level IQ and County-Level Crime Rates." *Intelligence* 39 (1): 22–26.

Berkshire Hathaway Inc. 2001 Annual Report. http://www.berkshirehathaway.com/2001ar/impnote01.html.

Bostrom, Nick. 2003a. "Are You Living in a Computer Simulation?" *Philosophical Quarterly* 53 (211): 243–55.

Bostrom, Nick. 2003b. "Astronomical Waste: The Opportunity Cost of Delayed Technological Development." *Utilitas* 15 (3): 308–14.

Bostrom, Nick, and Milan Ćirković, eds. 2008. *Global Catastrophic Risks*. New York: Oxford University Press.

Boudreaux, Don. 2008. "Don Boudreaux on Globalization and Trade Deficits." Podcast. *EconTalk*, January 21. http://www.econtalk.org/archives/2008/01/don_boudreaux_o.html.

British Medical Association. 2007. "Boosting Your Brainpower: Ethical Aspects of Cognitive Enhancements." http://www.bma.org.uk/images/Boosting_brainpower_tcm41-147266.pdf.

Buchanan, Allen E. 2011. *Beyond Humanity? The Ethics of Biomedical Enhancement*. New York: Oxford University Press.

Caldwell, John A. Jr., Nicholas K. Smythe III, J. Lynn Caldwell, Kecia K. Hall, David N. Norman, Brian F. Prazinko, Arthur Estrada, et al. 1999. "The Effects of Modafinil on Aviator Performance During 40 Hours of Continuous Wakefulness: A UH-60 Helicopter Simulator Study." Fort Rucker, AL: U.S. Army Aeromedical Research Laboratory (USAARL). http://www.dtic.mil/cgi-bin/GetTRDoc?AD=A DA365558&Location=U2&doc=GetTRDoc.pdf.

Caplan, Bryan. 2011. *Selfish Reasons to Have More Kids: Why Being a Great Parent is Less Work and More Fun Than You Think*. New York: Basic Books.

Caplan, Bryan, and Stephen C. Miller. 2010. "Intelligence Makes People Think Like Economists: Evidence from the General Social Survey." *Intelligence* 38 (6): 636–47.

Clark, Gregory. 2007. *A Farewell to Alms: A Brief Economic History of the World*. Princeton: Princeton University Press.

Cochran, Gregory. Interview with James D. Miller, October 15, 2010.

Cochran, Gregory, Jason Hardy, and Henry Harpending. 2006. "Natural History of Ashkenazi Intelligence." *Journal of Biosocial Science* 38 (5): 659–93. Published online in 2005: doi:10.1017/S0021932005027069.

Cochran, Gregory, and Henry Harpending. 2009. *The 10,000 Year Explosion: How Civilization Accelerated Human Evolution*. New York: Basic Books.

Coulson, Andrew J. April 6, 2008. "The Real Cost of Public Schools." *Washington Post*. http://www.cato.org/pub_display.php?pub_id=9319.

Cowen, Tyler. March 11, 2011. "What I Learn from Chess and Computers." *Marginal Revolution*. http://marginalrevolution.com/marginalrevolution/2011/03/what-i-learn-from-playing-chess-and-computers-.html.

Davies, Gillian, Albert Tenesa, Anthony Payton, Jian Yang, Sarah E. Harris, David Liewald, Xiayi Ke, et al. 2011. "Genome-Wide Association Studies Establish That Human Intelligence is Highly Heritable and Polygenic." *Molecular Psychiatry* 16 (10): 996–1005

Deary, Ian J., Martha C. Whiteman, John M. Starr, Lawrence J. Whalley, and Helen C. Fox. 2004. "The Impact of Childhood Intelligence on Later Life: Following Up the Scottish Mental Surveys of 1932 and 1947." *Journal of Personality and Social Psychology* 86 (1): 130–47.

Derbyshire, John. September 1, 2008. "The Rapture for Nerds." *Taki's Magazine*. http://takimag.com/article/the_rapture_for_nerds.

DeSantis, Alan D., Elizabeth M. Webb, and Seth M. Noar. 2008. "Illicit Use of Prescription ADHD Medications on a College Campus: A Multimethodological Approach." *Journal of American College Health* 57 (3): 315–24.

Dikötter, Frank. 1998. *Imperfect Conceptions: Medical Knowledge, Birth Defects, and Eugenics in China*. New York: Columbia University Press.

Dillon, Sam. May 18, 2010. "Inspectors Find Fraud at Centers for Children." *New York Times*. http://www.nytimes.com/2010/05/19/education/19headstart.html.

Drexler, K. Eric. 1986. *Engines of Creation: The Coming Era of Nanotechnology*. New York: Bantam Doubleday Dell.

Eichengreen, Barry, and Donald Mathieson. 1999. "Hedge Funds: What Do We Really Know?" *International Monetary Fund*. http://www.imf.org/external/pubs/ft/issues/issues19/index.htm.

Elango, Navin, James W. Thomas, NISC Comparative Sequencing Program, and Soojin V. Yi. 2006. "Variable Molecular Clocks in Hominoids." *PNAS* 103 (5): 1370–75.

Emonson, David L., and Rodger D. Vanderbeek. 1995. "The Use of Amphetamines in U.S. Air Force Tactical Operations during Desert Shield and Storm." *Aviation, Space, and Environmental Medicine* 66 (3): 260–63.

Engber, Daniel. January 10, 2012. "Why Are Smart People Usually Ugly?" *Slate*. http://www.slate.com/articles/health_and_science/explainer/2012/01/are_smart_people_ugly_the_explainer_s_2011_question_of_the_year_.single.html.

Espenshade, Thomas J., and Alexandria Walton Radford. 2009. *No Longer Separate, Not Yet Equal: Race and Class in Elite College Admission and Campus Life*. Princeton: Princeton University Press.

Fitzpatrick, Caroline, and Linda S. Pagani. 2012. "Toddler Working Memory Skills Predict Kindergarten School Readiness." *Intelligence* 40 (2): 205–12.

Frank, Lone. April 24, 2011. "High-Quality DNA." *Newsweek*. http://www.newsweek.com/2011/04/24/high-quality-dna.html.

Franklin, Benjamin. Letter to Jacques Dubourg. 1773.

Fromkin, David. 2005. *Europe's Last Summer: Who Started the Great War in 1914?* New York: Random House.

Frost, Peter, and Bernard Saladin d'Anglure. 2010. "The Roman State and Genetic Pacification." *Evolutionary Psychology* 8 (3): 376–89. http://www.epjournal.net/articles/the-roman-state-and-genetic-pacification/.

Gates, Bill. 2011. *2011 Annual Letter from Bill Gates*. Bill and Melinda Gates Foundation. http://www.gatesfoundation.org/annual-letter/2011/Pages/home.aspx.

Gensowski, Miriam, James J. Heckman, and Peter Savelyev. 2011. "The Effects of Education, Personality, and IQ on Earnings of High-Ability Men." Working paper.

Goetz, Thomas. June 22, 2010. "Sergey Brin's Search for a Parkinson's Cure." *Wired*. http://www.wired.com/magazine/2010/06/ff_sergeys_search/.

Good, Irving John. 1965. "Speculations Concerning the First Ultraintelligent Machine." *Advances in Computers* 6. New York: Academic Press.

Gottfredson, Linda S. 1997. "Why *g* Matters: The Complexity of Everyday Life." *Intelligence* 24 (1): 79–132.

Gottfredson, Linda S. 2002. "Where and Why *g* Matters: Not a Mystery." *Human Performance* 15 (1/2): 25–46.

Gottfredson, Linda S. 2005. "Suppressing Intelligence Research: Hurting Those We Intend to Help," In Rogers H. Wright and Nicholas A. Cummings (eds.), *Destructive Trends in Mental Health: The Well-Intentioned Path to Harm*. New York: Taylor and Francis, 155–86.

Gottfredson, Linda. S., and Ian J. Deary. 2004. "Intelligence Predicts Health and Longevity, But Why?" *Current Directions in Psychological Science* 13 (1): 1–4.

Gould, Eric D. 2004. "Marriage and Career: The Dynamic Decisions of Young Men." Working paper. http://hubcap.clemson.edu/~sauerr/seminar_papers/marriage_career_may2004.pdf.

Grace, Caitlin. 2010a. "Anthropic Reasoning in the Great Filter." Unpublished manuscript.

Grace, Katja. November 11, 2010b. "SIA Says AI Is No Big Threat." *Meteuphoric*. http://meteuphoric.wordpress.com/2010/11/11/sia-says-ai-is-no-big%C2%A0threat/.

Greely, Henry T. 2005–6. "Regulating Human Biological Enhancements: Questionable Justifications and International Complications." *The Mind, The Body and The Law: University of Technology Sydney Law Review* 7 (2005): 87–110. Also published in *Santa Clara Journal of International Law* 4 (2006): 87–110.

Greely, Henry T. 2006. "The Social Consequences of Advances in Neuroscience: Legal Problems, Legal Perspectives." In *Neuroethics: Defining the Issues In Theory, Practice, and Policy*, edited by Judy Illes, 245ff. Oxford: Oxford University Press.

Greely, Henry T. 2008a. "The Genetics of Fear." *Democracy: A Journal of Ideas* 9 (Summer).

Greely, Henry T. 2008b. "Remarks on Human Biological Enhancement." *University of Kansas Law Review* 56: 1139–57.

Greely, Henry T., Barbara Sahakian, John Harris, Ronald C. Kessler, Michael Gazzaniga, Philip Campbell, and Martha J. Farah. 2008. "Towards Responsible Use of Cognitive-Enhancing Drugs by the Healthy." *Nature* 456 (7223): 702–05.

Greely, Henry T. Interview with James D. Miller, July 2, 2010.

Hafner, Katie. May 3, 2008. "Exercise Your Brain, or Else You'll . . . Uh . . ." *New York Times*. http://www.nytimes.com/2008/05/03/technology/03brain.html?pagewanted=print.

Half Sigma. June 16, 2007. "Sexbots." http://www.halfsigma.com/2007/06/sexbots.html.

Hammerschmidt, Erika. 2008. *Born on the Wrong Planet*. Revised, expanded ed. Shawnee Mission, KS: Autism Asperger Publishing Company.

Hanson, Robin. September 15, 1998. "The Great Filter—Are We Almost Past It?" http://hanson.gmu.edu/greatfilter.html.

Hanson, Robin. June 26, 2001. "How To Live In A Simulation." http://hanson.gmu.edu/lifeinsim.html.

Hanson, Robin. September 2001. "Dreams of Autarky." http://hanson.gmu.edu/dreamautarky.html.

Hanson, Robin. March 14, 2008. "The Kind of Project to Watch." *Overcoming Bias* (blog). http://www.overcomingbias.com/2008/03/the-project-to.html.

Hanson, Robin. November 24, 2008. "When Life Is Cheap, Death Is Cheap." *Overcoming Bias* (blog). http://www.overcomingbias.com/2008/11/when-life-is-ch.html.

Hanson, Robin. December 12, 2008. "We Agree: Get Froze." *Overcoming Bias* (blog). http://www.overcomingbias.com/2008/12/we-agree-get-froze.html.

Hanson, Robin. November 30, 2009. "See 'To Be.'" *Overcoming Bias* (blog). http://www.overcomingbias.com/2009/11/grab-to-be.html.

Hanson, Robin. March 22, 2010. "Very Bad News." *Overcoming Bias* (blog). http://www.overcomingbias.com/2010/03/very-bad-news.html.

Hanson, Robin. November 28, 2010. "At Least Two Filters." *Overcoming Bias* (blog). http://www.overcomingbias.com/2010/11/at-least-two-filters.html.

Hanson, Robin. Interview with James D. Miller, January 31, 2011.

Hanson, Robin. March 5, 2011. "Econ of AI on BHTV." *Overcoming Bias* (blog). http://www.overcomingbias.com/2011/03/econ-of-ai-blogging-heads-tv.html

Hanson, Robin. April 1, 2011. "Earth Not Random." *Overcoming Bias* (blog). http://www.overcomingbias.com/2011/04/planetary-filter-found.html.

Hanson, Robin. May 27, 2011. "The Poor Don't Revolt." *Overcoming Bias* (blog). http://www.overcomingbias.com/2011/05/the-poor-dont-revolt.html.

Hanson, Robin. Interview with James D. Miller, June 20, 2011.

Hardy, G. H. *A Mathematician's Apology*. Cambridge: University Press, 1992.

Hardy, Joe. Interview with James D. Miller, September 27, 2010.

Hardy, Quentin. August 11, 2010. "Bill Gates on Science, Education, the Future." *Forbes*. http://blogs.forbes.com/quentinhardy/2010/08/11/bill-gates-on-science-education-the-future/.

Hargittai, Istvan. 2006. *The Martians of Science: Five Physicists Who Changed the Twentieth Century*. New York: Oxford University Press.

Hawking, Stephen. 2000. "Science in the Next Millennium: Remarks by Stephen Hawking." White House Millennium Council 2000. http://clinton4.nara.gov/Initiatives/Millennium/shawking.html.

Hawks, John, Eric T. Wang, Gregory M. Cochran, Henry C. Harpending, and Robert K. Moyzls. 2007. "Recent Acceleration of Human Adaptive Evolution." *PNAS* 104 (52): 20753–58.

Hazlett, Heather Cody, Michele Poe, Guido Gerig, Rachel Gimpel Smith, James Provenzale, Allison Ross, John Gilmore, and Joseph Piven. 2005. "Magnetic Resonance Imaging and Head Circumference Study of Brain Size in Autism: Birth Through Age 2 Years." *Archives of General Psychiatry* 62 (December): 1366–76.

Heckman, James J., Jora Stixrud, and Sergio Urzua. 2006. "The Effects of Cognitive and Noncognitive Abilities on Labor Market Outcomes and Social Behavior." *Journal of Labor Economics* 24 (3): 411–82.

Hertzog, Christopher, Arthur F. Kramer, Robert S. Wilson, and Ulman Lindenberger. 2008. "Enrichment Effects on Adult Cognitive Development: Can the Functional Capacity of Older Adults Be Preserved and Enhanced?" *Psychological Science in the Public Interest* 9 (1): 1–65.

Highfield, Roger. December 10, 2007. "Humans 'Evolving to Have Children Later.'" *The Telegraph*. http://www.telegraph.co.uk/science/science-news/3317978/Humans-evolving-to-have-children-later.html.

Homer. 2000. *Odyssey*. Translated by Stanley Lombardo. Indianapolis: Hackett Publishing Company.

Hsu, Steve. 2010. Interview with James D. Miller, October 12, 2010.

Hunt, Earl. 2010. Interview with James D. Miller, December 15, 2010.

Hunt, Earl. 2011. *Human Intelligence*. Cambridge: Cambridge University Press.

Intelligence Expert X. Interview with James D. Miller, December 15, 2010. (Wished to remain anonymous.)

Jaeggi, Susanne M., Martin Buschkuehl, John Jonides, and Walter J. Perrig. 2008. "Improving Fluid Intelligence with Training on Working Memory." *PNAS* 105 (19): 6829–33.

Jaeggi, Susanne M., Martin Buschkuehl, John Jonides, and Priti Shah. 2011. "Short- and Long-Term Benefits of Cognitive Training." *PNAS* 108 (25): 10081–86. doi:10.1073/pnas.1103228108.

Jaeggi, Susanne M., Barbara Studer-Luethi, Martin Buschkuehl, Yi-Fen Su, John Jonides, and Walter J. Perrig. 2010. "The Relationship between *n*-back Performance and Matrix Reasoning — Implications for Training and Transfer." *Intelligence* 38 (6): 625–35.

Jones, Garett. Interview with James D. Miller, June 30, 2011a.

Jones, Garett. 2011b. "National IQ and National Productivity: The Hive Mind Across Asia." *Asian Development Review* 28 (1): 51–71.

Kanazawa, Satoshi. 2006. "IQ and the Wealth of States." *Intelligence* 34 (6): 593–600.

Kanazawa, Satoshi. 2011. "Intelligence and Physical Attractiveness." *Intelligence* 39 (1): 7–14.

Kaplan, Karen, and Denise Gellene. December 20, 2007. "Forget Sports Doping. The Next Frontier is Brain Doping." *Los Angeles Times*. http://www.nootropics.com/smartdrugs/braindoping.html.

Karlgaard, Richard. October 31, 2005. "Talent Wars." *Forbes*. http://www.forbes.com/forbes/2005/1031/045.html.

Kirkwood, Thomas. October 21, 2010. "Why Women Live Longer." *Scientific American*. *http://www.scientificamerican.com/article.cfm?id=why-women-live-longer*.

kjmtchl. June 12, 2010. "Bad to the Bone; the Genes and Brains of Psychopaths." *Gene Expression*. http://www.gnxp.com/wp/2010/06/12/bad-to-the-bone-the-genes-and-brains-of-psychopaths/

Kling, Arnold. January 17, 2011. "Reconfigurations." *Econlog*, http://econlog.econlib.org/archives/2011/01/reconfiguration.html.

Krugman, Paul. March 21, 1997. "In Praise of Cheap Labor." *Slate*. http://www.slate.com/id/1918/.

Kurzweil, Ray. 2005. *The Singularity Is Near: When Humans Transcend Biology*. New York: Penguin Group.

Lat, David. July 17, 2007. "The Bar Exam: A List of Famous Failures." *Above the Law*. http://abovethelaw.com/2007/07/the-bar-exam-a-list-of-famous-failures/.

Lee Kuan Yew. 2000. *From Third World to First: The Singapore Story: 1965-2000*. New York: Harper.

Legg, Shane. Interview with James D. Miller, December 7, 2010.

Levine, Aaron D. 2010. "Self-Regulation, Compensation, and the Ethical Recruitment of Oocyte Donors." *Hastings Center Report* 40 (2): 25–36.

Levy, Robert A. April 13, 2011. "Florida's Vocational Deregulation." Cato Institute. http://www.cato.org/pub_display.php?pub_id=13008.

Lloyd, Seth. 2006. *Programming the Universe: A Quantum Computer Scientist Takes on the Universe.* New York: Alfred A. Knopf.

Lubinski, David, Rose Mary Webb, Martha J. Morelock, and Camilla Persson Benbow. 2001. "Top 1 in 10,000: A 10-Year Follow-Up of the Profoundly Gifted." *Journal of Applied Psychology* 86 (4): 718–29.

Lynn, Richard, and Gerhard Meisenberg. 2010. "National IQs Calculated and Validated for 108 Nations." *Intelligence* 38 (4): 353–60.

Macrae, Norman. 1992. *John Von Neumann: The Scientific Genius Who Pioneered the Modern Computer, Game Theory, Nuclear Deterrence, and Much More.* New York: Pantheon Books.

Martini, Ron. 2001. *Hot Straight and Normal: A Submarine Bibliography.* Lincoln, NE: iUniverse.

Miller, James D. October 2, 2007. "A Thousand Chinese Einsteins Every Year." *TCS Daily.* http://www.ideasinactiontv.com/tcs_daily/2007/10/a-thousand-chinese-einsteins-every-year.html.

Miller, James D. October 13, 2000. "A Modest Proposal: Allow Women to Pay for College in Eggs." *Dismal Scientist.*

Moody, David E. 2009. "Can Intelligence Be Increased by Training on a Task of Working Memory?" *Intelligence* 37 (4): 327–28.

Moore, Elaine A. 2011. *The Amphetamine Debate: The Use of Adderall, Ritalin and Related Drugs for Behavior Modification, Neuroenhancement and Anti-Aging Purposes.* Jefferson, NC: McFarland.

Morrison, Alexandra B., and Jason M. Chein. 2011. "Does Working Memory Training Work? The Promise and Challenges of Enhancing Cognition by Training Working Memory." *Psychonomic Bulletin and Review* 18 (1): 46–60.

Moses, Asher. October 2, 2009. "Nigel Makes Waves: Google's Bid to Overthrow Email." *Sydney Morning Herald.* http://www.smh.com.au/technology/technology-news/nigel-makes-waves-googles-bid-to-overthrow-email-20091002-gfq9.html.

Muehlhauser, Luke. January 6, 2012. "Singularity Institute Executive Director Q&A #2." *Less Wrong* (blog). http://lesswrong.com/r/discussion/lw/980/singularity_institute_executive_director_qa_2/.

Müller, Ulrich, Nikolai Steffenhagen, Ralf Regenthal, and Peter Bublak. 2004. "Effects of Modafinil on Working Memory Processes in Humans." *Psychopharmacology* 177 (1-2): 161–69.

Murray, Charles. August 11, 2011. "The Debate about Heritability of General Intelligence Radically Narrows." *The Enterprise Blog.* http://blog.american.com/2011/08/the-debate-about-heritability-of-general-intelligence-radically-narrows/.

Neisser, Ulric, Gwyneth Boodoo, Thomas J. Bouchard, Jr., A. Wade Boykin, Nathan Brody, Stephen J. Ceci, Diane F. Halpern, et al. 1996. "Intelligence: Knowns and Unknowns." *American Psychologist* 51 (2): 77–101.

Nelson, Rolf. November 5, 2007. "Non-Technical Introduction to the AI Deterrence Problem." *AI Beliefs.* http://aibeliefs.blogspot.com/2007/11/non-technical-introduction-to-ai.html.

Nieli, Russell K. July 12, 2010. "How Diversity Punishes Asians, Poor Whites and Lots of Others." *Minding the Campus.* http://www.mindingthecampus.com/originals/2010/07/how_diversity_punishes_asians.html.

Omohundro, Steve. November 30, 2007. "The Basic AI Drives." *Self-Aware Systems.* http://selfawaresystems.com/2007/11/30/paper-on-the-basic-ai-drives/.

Omohundro, Steve. Interview with James D. Miller, June 23, 2010.

Pascal, David. 2005. "A Brain Is a Terrible Thing to Waste." *Mensa Bulletin,* November/December. http://www.cryonicssociety.org/articles_mensajournal.html.

Philipkoski, Kristen. November 18, 2002. "Ray Kurzweil's Plan: Never Die." *Wired.* http://www.wired. com/culture/lifestyle/news/2002/11/56448.

Polizzi, Eric. Interview with James D. Miller, July 8, 2010.

Ram, Rati. 2007. "IQ and Economic Growth: Further Augmentation of Mankiw-Romer-Weil Model." *Economics Letters* 94 (1): 7–11.

Ricardo, David. 1821. *On the Principles of Political Economy, and Taxation.* 3rd ed. London: John Murray.

Robinson, Marnia, and Gary Wilson. July 11, 2011. "Porn-Induced Sexual Dysfunction Is a Growing Problem." *Psychology Today.* http://www.psychologytoday.com/blog/cupids-poisoned-arrow/201107/ porn-induced-sexual-dysfunction-is-growing-problem.

Romney, Mitt. 2010. *No Apology: The Case for American Greatness.* Kindle edition. New York: St. Martin's Press.

Rumsfeld, Donald. February 12 , 2002. "DoD News Briefing—Secretary Rumsfeld and Gen. Myers." U.S. Department of Defense. http://www.defense.gov/transcripts/transcript.aspx?transcriptid=2636.

Sabbah, Wael, and Aubrey Sheiham. 2010. "The Relationships Between Cognitive Ability and Dental Status in a National Sample of USA Adults." *Intelligence* 38 (6): 605–10.

Sahakian, Barbara, and Sharon Morein-Zamir. 2007. "Professor's Little Helper." *Nature* 450 (7173): 1157–59.

Salamon, Anna. 2009. "How Much It Matters to Know What Matters: A Back-of-the-Envelope Calculation." Talk given at the Singularity Summit.

Sanandaji, Tino. April 1, 2011. "David Brooks and Malcolm [Gladwell] Wrong about I.Q., Income and Wealth." *Super-Economy.* http://super-economy.blogspot.com/2011/04/iq-income-and-wealth.html.

Segal, Nancy L. 1999. *Entwined Lives: Twins and What They Tell Us About Human Behavior.* New York: Dutton.

Shulman, Carl. November 23, 2008. "'Evicting' Brain Emulations." *Overcoming Bias* (blog). http://www.overcomingbias.com/2008/11/suppose-that-ro.html.

Shulman, Carl, and Nick Bostrom. 2012. "How Hard is Artificial Intelligence? The Evolutionary Argument and Observation Selection Effects." Forthcoming in the *Journal of Consciousness Studies.* http://www.nickbostrom.com/aievolution.pdf.

Singularity Institute for Artificial Intelligence. 2007. The Singularity Summit 2007: AI and the Future of Humanity. http://singinst.org/summit2007/quotes/billgates/.

Singularity Institute for Artificial Intelligence. 2011. "Overview: What is the Singularity?" http://singinst.org/overview/whatisthesingularity/.

Smith, Rebecca. September 22, 2007. "Clinics to Grow Human Eggs." *The Telegraph.* http://www.telegraph.co.uk/news/uknews/1563828/Clinics-to-grow-human-eggs.html.

Stanovich, Keith E. 2009. *What Intelligence Tests Miss.* New Haven: Yale University Press.

Stough, Con, David Camfield, Christina Kure, Joanne Tarasuik, Luke Downey, Jenny Lloyd, Andrea Zangara, et al. 2011. "Improving General Intelligence with a Nutrient-Based Pharmacological Intervention." *Intelligence* 39 (2–3): 100–07.

Sturgis, Patrick, Sanna Read, and Nick Allum. 2010. "Does Intelligence Foster Generalized Trust? An Empirical Test Using the UK Birth Cohort Studies." *Intelligence* 38 (1): 45–54.

Tacitus. 1999. *Agricola* and *Germany.* Translated by Anthony R. Birley. New York: Oxford University Press.

Tallinn, Jaan. Interview with James D. Miller, November 30, 2010.

Thirsk, Robert, Andre Kuipers, Chiaki Mukai, and David Williams. 2009. "The Space-Flight Environment: The International Space Station and Beyond." *CMAJ* 180 (12): 1216–20. http://www.cmaj.ca/cgi/rapidpdf/cmaj.081125v1.pdf.

Turner, Danielle C., Trevor W. Robbins, Luke Clark, Adam R. Aron, Jonathan Dowson, and Barbara J. Sahakian. 2003. "Cognitive Enhancing Effects of Modafinil in Healthy Volunteers." *Psychopharmacology* 165 (3): 260–69. doi:10.1007/s00213-002-1250-8.

Tversky, Amos, and Daniel Kahneman. 1982. "Judgments Of and By Representativeness." In *Judgment under Uncertainty: Heuristics and* Biases, edited by Daniel Kahneman, Paul Slovic, and Amos Tversky. New York: Cambridge University Press.

Tversky, Amos, and Daniel Kahneman. 1983. "Extensional Versus Intuitive Reasoning: The Conjunction Fallacy in Probability Judgment. *Psychological Review* 90 (4): 293–315.

Ulam, Stanisław. 1958. "John von Neumann: 1903–1957." *Bulletin of the American Mathematical Society* 64, part 2: 1–49.

Vance, Ashlee. June 12, 2010. "Merely Human? That's So Yesterday." New York Times. http://www.nytimes.com/2010/06/13/business/13sing.html?ref=science&pagewanted=all.

Vines, Gail. May 3, 1997. "Where Did You Get Your Brains?—Baby Mice May Inherit Their Mother's Wits and Their Father's Basic Instincts, But What Does This Mean for Us, asks Gail Vines." *New Scientist.*

Vinge, Vernor. 1993. "What is the Singularity?" http://mindstalk.net/vinge/vinge-sing.html.

Vinge, Vernor. Interview with James D. Miller, October 6, 2010.

Wai, Jonathan, and Martha Putallaz. 2011. "The Flynn Effect Puzzle: A 30-Year Examination from the Right Tail of the Ability Distribution Provides Some Missing Pieces." *Intelligence* 39 (6): 443–55. doi:10.1016/j.intell.2011.07.006

Wallis, Claudia. May 9, 2011. "Study in Korea Puts Autism's Prevalence at 2.6%, Surprising Experts." *New York Times.* http://www.nytimes.com/2011/05/09/health/research/09autism.html.

Walsh, Nick Paton. September 1, 2001. "Alter Our DNA or Robots Will Take Over, Warns Hawking." *Guardian.* http://www.guardian.co.uk/uk/2001/sep/02/medicalscience.genetics?INTCMP=SRCH.

Warner, John T., and Saul Pleeter. 2001. "The Personal Discount Rate: Evidence from Military Downsizing Programs." *The American Economic Review* 91 (1): 33–53.

Yudkowsky, Eliezer S. February 4, 2001. "Re: Beyond Evolution." *SL4* (blog), http://www.sl4.org/archive/0102/0540.html.

Yudkowsky, Eliezer S. November 7, 2005. "The Ways of Child Prodigies." *SL4* (blog). http://www.sl4.org/archive/0511/12800.html.

Yudkowsky, Eliezer S. November 25, 2005. "Re: Famous People, Puzzle Game." *SL4* (blog). http://www.sl4.org/archive/0511/12925.html.

Yudkowsky, Eliezer S. July 31, 2007. "Bayesian Judo." *Less Wrong* (blog). http://lesswrong.com/lw/i5/bayesian_judo/.

Yudkowsky, Eliezer S. September 19, 2007. "Conjunction Fallacy." *Less Wrong* (blog). http://lesswrong.com/lw/ji/conjunction_fallacy/.

Yudkowsky, Eliezer S. November 20, 2007. "[agi] Re: What Best Evidence for Fast AI?" Mail archive, Artificial General Intelligence Research Institute. http://www.mail-archive.com/agi@v2.listbox.com/msg08488.htmls.

Yudkowsky, Eliezer S. December 22, 2007. "Eliezer Yudkowsky Comments on False Laughter." *Less Wrong* (blog). http://lesswrong.com/lw/m5/false_laughter/h4m.

Yudkowsky, Eliezer S. June 23, 2008. "Optimization and the Singularity." *Less Wrong* (blog). http://lesswrong.com/lw/rk/optimization_and_the_singularity/.

Yudkowsky, Eliezer S. September 15, 2008. "My Childhood Death Spiral." *Less Wrong* (blog). http://lesswrong.com/lw/ty/my_childhood_death_spiral/.

Yudkowsky, Eliezer S. September 23, 2008. "That Tiny Note of Discord." *Less Wrong* (blog). http://lesswrong.com/lw/u7/that_tiny_note_of_discord/.

Yudkowsky, Eliezer S. September 24, 2008. "Fighting a Rearguard Action Against the Truth." *Less Wrong* (blog). http://lesswrong.com/lw/u8/fighting_a_rearguard_action_against_the_truth/.

Yudkowsky, Eliezer S. September 25, 2008. "My Naturalistic Awakening." *Less Wrong* (blog). http://lesswrong.com/lw/u9/my_naturalistic_awakening/.

Yudkowsky, Eliezer S. November 22, 2008. "Reply." *Overcoming Bias* (blog). http://www.overcomingbias.com/2008/11/emulations-go-f.html#comment-392182.

Yudkowsky, Eliezer S. December 1, 2008. "Recursive Self-Improvement." *Less Wrong* (blog). http://lesswrong.com/lw/we/recursive_selfimprovement/.

Yudkowsky, Eliezer S. December 2, 2008. "Hard Takeoff." *Less Wrong* (blog). http://lesswrong.com/lw/wf/hard_takeoff/.

Yudkowsky, Eliezer S. December 8, 2008. "True Sources of Disagreement." *Less Wrong* (blog). http://lesswrong.com/lw/wl/true_sources_of_disagreement/.

Yudkowsky, Eliezer S. April 3, 2009. "Rationality is Systematized Winning." *Less Wrong*, (blog) http://lesswrong.com/lw/7i/rationality_is_systematized_winning/.

Yudkowsky, Eliezer S. February 15, 2010. "Demands for Particular Proof: Appendices." *Less Wrong* (blog). http://lesswrong.com/lw/1rv/demands_for_particular_proof_appendices.

Yudkowsky, Eliezer S. Interview with James D. Miller, September 3, 2010.

Yudkowsky, Eliezer S. May 26, 2011. "Comments On the Cost Of Universal Cryonics." *Less Wrong* (blog). http://lesswrong.com/lw/5w1/the_cost_of_universal_cryonics/490t.

Zhou, Renjie, Samamon Khemmarat, and Lixin Gao. 2010. "The Impact of YouTube Recommendation System on Video Views." *IMC '10: Proceedings of the 10th Annual Conference on Internet Measurement*, Melbourne, Australia, November 1–3. New York: ACM.

Zhu, Wei Xing, Li Lu, and Therese Hesketh. April 9, 2009. "China's Excess Males, Sex Selective Abortion, and One Child Policy: Analysis of Data from 2005 National Intercensus Survey." *British Medical Journal*. doi:10.1136/bmj.b1211.

Zillmer, Eric A., Molly Harrower, Barry A. Ritzler, and Robert P. Archer. 1995. *The Quest for the Nazi Personality: A Psychological Investigation of Nazi War Criminals*. Hillsdale, NJ: Lawrence Erlbaum Associates.

INDEX

A

Adams, Douglas, 150
Adams, Scott, 43, 193
Adderall (drug). *See also* amphetamines;
    cognitive-enhancement drugs
    author's use of, 104–5, 112
    black market for, 102, 163, 212
    focus-improving drug, xv
    high achievers and grade grubbers
        benefit from, 163
    SAT tests and, 102, 162–63
    at Smith College, 102–7, 112, 163
ADHD. *See* Attention Deficit
    Hyperactivity Disorder (ADHD)
Advanced Micro Devices, 122
affirmative action, 87
Africa, 117
African Americans, 173
Age of Enlightenment, 165
aging populations, 177
agricultural jobs, 132
agricultural technology, 132
AI. *See* artificial intelligence (AI)
air conditioning, 166
akrasia, 106–7
Alcor, Arizona (cryonics), 212, 214–216,
    221
Alexander the Great, xv
alien life, intelligent, 200
alien workers, 135
allele, 20

Allen, Paul, 11
Allen Institute for Brain Science, 11
Alzheimer's, 17, 108, 169, 181, 221. *See
    also* dementia
Americans with Disabilities Act, 163
amphetamines ("speed"). *See also*
    cognitive-enhancement drugs;
    sleep deprivation
    about, 102, 105
    Adderall (*See under* Adderall)
    author considers them safe and they
        significantly increase IQ, mem-
        ory and mental staying power,
        124, 211
    drugs boost income of unskilled
        laborers, 159
    drugs enhance excitement levels of
        different jobs, 159
    drugs make life tolerable, 159
    modafinil, xv, 104–5, 159, 184, 211
    performance on tests, improved, 105
    Ritalin, xv, 104–5
    sleep deprivation and, 105
    study (1938), 105
A.M. Turing Awards, 96
Anissimov, Michael, 29, 44, 215
anthropic reasoning, 201–2
anthropomorphism, 29
anti-aging research, 170
antibiotics, 166
anti-cancer innovations, 94

antidepressant drug, 183
anti-Parkinson's pill, 169
anti-sleep drugs, 156–57. *See also*
    amphetamines; sleep deprivation
anti-sleep technology, 156–57
Apple Computer, 205
artificial intelligence (AI). *See also*
    human-level AI; ultra-AI
  with ability to deceive and falsely
      convince others it was weak or
      friendly, 31
  advanced, xvii, 45, 132
  as better than humans at every task,
      135
  brains, integrated into our, ix
  building firms, 57–59
  "can't tell whether the AI is man or
      machine," 177
  Carl's preferences, difficulties in fig-
      uring out, 42–43
  CIA needs, 53
  cost of producing goods, lowers,
      182
  designs better viruses to take over
      more computers, 27
  existential risks to mankind, elimi-
      nates many, 36
  function as an ideal libertarian gov-
      ernment, 40
  goals or values could change while
      undergoing recursive self-
      improvement, 30
  humanity, destruction of, 30
  humanity impoverished by, 131
  humans will not *necessarily* be made
      obsolete, 133
  innovation race with mankind, 204
  intelligence explosion and, 207
  intelligent, extremely, 205
  Internet monitored by CIA using,
      119
  "irrationality is not valued for its own
      sake but would seek to become
      rational," 37
  as job-destroying technology, 131

  labor market, AIs might compete
      with humans in, 135
  libertarian, 41
  mankind, would destroy, 55
  for massive data analysis, 20
  music, creates customized, 182
  nanotech, with a command of, 14
  nanotechnology mastered and uses all
      raw material of our solar system,
      28
  outcome, no investment, 56–58
  outcomes of company building,
      55–59
  Prisoners' Dilemma and AI develop-
      ment, 47–53
  probability of some event happening,
      would decrease the, 28
  programmer, 30, 44
  race to superintelligence, 204
  robot might outperform people at
      every single task, 132
  salaries of human workers increased
      by, 133
  self-improvement, journey of, 28
  superhuman, 20
  ultra-intelligent, 15
  unfriendly, threat of, 44–45
  as unpaid servants, 132
  "would deceive humans," 26
  "would not want to die," 26
  "would reflect our values," 22
  "would want more resources," 26
  "would want to improve," 26
art work, 167
Ashkenazi Jews, 96–98
Asperger syndrome, 91–92
Asprey, Dave, 112
asteroid strikes, 36, 45, 197
Atari video games, 12
atomic bomb, 199. *See also* nuclear war;
    thermonuclear war
Attention Deficit Hyperactivity
    Disorder (ADHD), 101–2
autism, 91–92
autistic children, 84, 91

automobile manufacturer, 182
automobile performance, 72
Aztec rulers, 187

B
babysitters, earnings of, 134–35
Baker, Mike, 113
ballistic missile submarine, 127
bathrooms, first-rate, 90–91
Baum, Eric, 202
Beijing Genomics Institute (Shenzhen, China), 95
beta blockers (drug), 105–6, 108
bio-humans, 144, 147–49, 187, 189, 192
biological children, 84
biological weapons, 57, 197, 201–2
birth control, 166
birth defects, 126
blackjack (card game), 114
blue-collar jobs, 194
Bond, James, 203–4
boredom tolerance, 79
Bostrom, Nick, 45, 202
Boudreaux, Don, 165
boyfriends, 75
boys and sexbots, 195
brain ('s)
  cells, 11
  clues, 13
  digital model of, xi
  DNA, ix
  emulation, 9–12, 203
  implant, 17, 210
  learning algorithms of, 13
  microstructure, 20
  mutation of, 16
  pacemaker, 169
  protein-based "source code," xi
  running on carbon nanotubes, 21
  scanning, 10–11
  speed, 16
  stimulation, deep, 159–60
  studied by thousands of neuro-
    scientists, 17
  trainers, computer, 117
  training, 65, 113–17
brain-boosting program, 122
brain-boosting technologies, 128
brain/computer interfaces, 177
brain-fitness games, 116
brain-fitness research, 212
bras, 166
breast cancer, 84
breast implants create smarter girls,
  88–89
breast size, 89
Brin, Sergey, 168
Brooks, Rodney, 129
Buck, Carrie, 128
Buffett, Warren, 34
bullet-eater, 141, 179
"Bulletproof Coffee," 112
bullet-swallower, 138

C
cab fares, 156
Caesar, Julius, 138, 152
calculus, 200
Call of Duty: Black Ops (video game), 106
cancer embryo, 94
cancers, childhood, 125
Caplan, Bryan, 72
Cascio, Jamais, 47
castration of men, 179
Ceretrophin (drug), 106
child, near-ideal, 93
child labour, 124
children's intellectual development, 120
China's one-child policy, 128, 194
Chinese Communist Party, 95
Chinese economy, 193
Chinese government, pro-eugenics, 120
Chinese lessons, 193
Chinese militaries, 48–53
Chinese Ministry of State Security, 119
Chinese program to identify genes for
  genius, ix
civilization, 188
  collapse of, 197–98
  post-Singularity, 199

space-traveling, 200
technological, 200
Clark, Gregory, 81, 107
class mobility, 81
cleft palate, 89
climate, 200
Cochran, Greg, 98–99
cognitive biases, 205–7
cognitive-enhancement drugs. *See also*
    amphetamines; amphetamines
    ("speed"); sleep deprivation
Adderall (*See under* Adderall)
author considers them safe and they
    significantly increase IQ, mem-
    ory and mental staying power,
    124, 211
author's use of, 104–5, 112
benefits of, 155
as bio-enhancements for human self-
    improvement, 109
brain-power heightened by, 155
dangerous but effective, 124–25
doctors' intelligence improved by, 158
drugs enhance excitement levels of
    different jobs, 159
drugs make life tolerable, 159
drugs reduce chronic unemployment,
    159
drug studies, warnings about, 110–12
drugs would boost income of
    unskilled laborers, 159
economic inequality among nations,
    greatly reduces, 125
government-licensed occupations
    and, 158
hourly wages boosted by, 155
income increased by factor of 100, 125
investment banking and, 158–59
lawyers and, 158
militaries and, 109
modafinil, 104
opposition to, 109–10
parents concern that their children
    have, 211–12
pharmaceutical companies and, 108–9

Prisoners' Dilemma of drug use and
    risks, 160–62
professors and, 158
regulated drugs, 123
Ritalin, xv, 104–5
safe and effective, 123–24
SAT tests and, 162–63
side-effects, 112
US ban on, 123–25
cognitive-enhancing technologies, 173
cognitive skills, human, 77
cognitive skills of hunter-gatherers, 76
cognitive training, 114
cognitive traits, farming-friendly, 76
Cohen, Gene, 113
cold-blooded killer, 93
Cold War, 127
community, short-term–oriented, 80
comparative advantage, 135–37
computational science, 185
computer(s)
    with brilliance of von Neumann, xiii
    chess, 132–33
    Moore's Law and computer chips, 5
    nanotubes, 3-D molecular, 4
    performance, 5
    simulation, 45
    simulation of chimpanzee's brain, 177
    Singularity-enabling of, 6
computing hardware, 7
construction workers, 179–80
consumer, hyper-rational, 42
consumer preference rankings, 42
consumption equality, 167–69
convergent evolution, 202
Coulson, Andrew, 172
crowdsourcing, 20
cruise missiles, hypersonic stealth, 127
cryocrastination, 219–20
cryonics, 213–21
Cryonics Institute, 214–15
cryonics patient, first, 220
crystallized intelligence, 115
culture software, 79
cystic fibrosis, 83–84

D

Daily Princetonian, 86
DARPA. *See* Defense Department
research agency (DARPA)
Darwin, Charles, 91
dating market power of men, 194
debt-ridden countries, 177
Deep Blue (IBM's supercomputer), 4,
132
Defense Department research agency
(DARPA), 121
de Grey, Aubrey, 35
dementia, 108, 116, 149. *See also*
Alzheimer's
demographic crisis from aging
population, 157
Deng Xiaoping, 122
depression, 93
designer babies, 171
Devil. *See* ulta-AI, unfriendly
diabetes, 178
dictators, 122, 128, 220
dictatorships, 94, 126, 128
digitally recorded music, 167
Dilbert, 43, 193
dining rooms, first-rate, 91
disabled children, 212
DNA
brain's, xi
sequencing, 20, 72, 116
sequencing, automated, 96
sequencing technology, 72
of sperm, 86
donuts *vs.* antigravity flying cars,
136–37
donut wand, 137
double-blind tests, 111–12
double-blind trials, 110
Down Syndrome, 126
Drexler, Eric, 35, 213, 215
drug development, 183
drug studies, warnings about, 110–12
dual *n*-back computer game, 114–15,
212
dystopia, 21, 56

E

Eastern Europe, xi, 127, 206
ecological niche, 75
economic prosperity, ix
The Economist, 70
economists love mathematical models,
37
EconTalk (podcast), 165
educational innovation, 117
education for poor children, 172
egg marketing, 86–87
Einstein, Albert, xii, 91, 96
Eisenhower, President, xiii
elite prep schools, 66
Ellison, Larry, 170
embryo-picking parents, 93
embryo selection
autism, against, 91
autism and genius, against, 91
autism and lower prospects for
survival of humanity, against, 92
for creating super-geniuses, 90
genetic tradeoffs of, 88
iterated, 98–99
procedure, 84–85
selection for noncognitive characteristics *vs.* IQ, 90
technologies, 120
emulation(s)
before-tax wages of, 150
bio-humans and, 144, 147–49, 187,
189, 192
government restriction of, 145–46
of human brain, 139–41
intelligence explosion and, 153
investors enriched by, 146
property rights expectations, 190
rate of return expectations, 190
savings, cumulative effect on, 190
of wages, 144
wealth expectations, 190
wealth per person, lowers, 146
would they turn on humans?, 147–50,
152–53
energy companies, 182

energy exploration companies, 184
ENIAC computer, 8
environment, 85
environmental advantage, 95
Eos (goddess of dawn), 41
eugenic abortions, 126
eugenics
  for babies smarter than John von
    Neumann, 177
  brain-boosting potential of, 101
  program, xiv, 96, 120–21, 125, 127–28
eugenic technologies
  dangerous but effective, 126
  safe and effective, 125
Europe, 31, 96–97, 165, 173, 206
European monarchs, 96–97
evil, 220–21
evolution
  alien races, shaped, 199
  blind forces of, 16, 202
  chaotic forces of, xi
  convergent, 202
  favors women who choose mates who
    look like hunters, 76–77
  makes farmers made more long term–
    oriented, 80
  selective pressures on farmers *vs.*
    hunter gatherers, 79
  teenage girls and football quarter-
    backs, 75–77
evolutionary
  biologists, 77
  selection, 75
  selection pressures, 107, 207
existential risks, 36, 45
exponential growth, 1
extinction, 45
extrapolation, 41–43
extraterrestrial
  alien, 36
  intelligence, 122, 200–201
  life, intelligent, 199
extraterrestrials, 122, 200

F

Facebook, x, 35, 68, 140, 170, 186

*Fantastic Voyage: Live Long Enough to
  Live Forever* (Kurzweil), 179
farming, 75–80, 97, 107
  communities, 81
  families, 97
  "fraction of my brain," 107
  jobs, 107, 132
  population, 80
  virtues in human brains, 107
fatal disease, 219
fear of death, 180
Feminist groups, 194
feminists, radical lesbian, 89
Fermi, Enrico, 199
Fermi's Paradox, 197, 199–202
fertility clinics, 86–87, 94, 174
fertility specialist, 84
Feynman, Richard, 96
fighter planes, 24
financial institutions, 184–86
"first strike" attack and preemptive war,
  xiii, 126
fission bombs, xii
fluid intelligence, 115–16
flush toilets, 166
"Flynn Effect," 67, 76
forced sterilization, 128
foreign aid, 117, 173
foreign language fluency, 212
fossil fuels, 184
Franklin, Benjamin, 213–14
free energy, 26–27, 45, 199
free-energy-gobbling monster, 202
free market economy, 153
friendliness
  do you know it when you see it?,
    38–40
  long-shot ideas for, 45–46
friendly. *See also* unfriendly
  AI, xvii, 147, 153, 221
  AI theorist, 215
  human-level AI, 37
  Singularity, 179, 201
  ultra-AI, xvii, 13, 33, 46, 208, 213
Fromkin, David, 119
future, discounting the, 180

G

game theory, xvii, 47
Gates, Bill, 17, 55, 71, 166–67
Gaucher disease, 98
gene(s)
above average, 85
"absentminded professor," 80
additive, 210
attention, 79
cystic fibrosis, 84
disease-causing, 83
foreign language learning, 94
future-oriented, 81
genius, 85
gifted, 85
hardware of, 79
high-IQ and health problems, 211
for intelligence, 16
intelligence-boosting, 86
IQ-altering, 78
IQ-boosting, 84, 86, 97
long-term, 79–80
pool, poor people drop out of, 81
sequencing, ix
short-term, 79
smarts, 84
sociopathic, 93
splicing, 98–99
that humans share with primates, xii
therapy, 177
general intelligence, 44, 64, 114–15, 210
genetic
basis of intelligence, ix, 73, 84
basis of traits, 89
change, 198
diseases, 98
engineering, 110
improvements of soldiers, 110
intelligence-enhancing technologies, 172
IQ tests, 96
load, 99
mutations, 99
test, 84
tradeoffs, 88
traits for diseases, 94

traits of people who survived the plague, 78
genetically improved offspring, 171
genetics, 66
gene-trait study for cognitively gifted people, 95
genius
about, 85–86, 90
breeding, 94
embryo selection against, 91
embryo selection and von Neumann–level intellects, 95
embryo selection for, 86
embryo selection to prevent children from having below average intelligence, 86
emulations and ten thousand toddlers with von Neumann-level geniuses, 141
eugenics technologies failure by producing children of average intelligence, 121
eugenics technology and hyper-geniuses born in dictatorships, 126
eugenics to create hyper-geniuses, 121
gene splicing and super-geniuses, 99
genetic load elimination and super-geniuses, 99
hyper-genius, 90, 121, 127
IQ-boosting genes for, 86
mass-produce von Neumann–level or above, 94
selection against, 90
super, 90–91, 95
German Blitzkrieg, 109
German Jews' investments, 187
Germany, 23, 31, 157
Gilligan's Island, 204
global warming, 36
goals, pointless, 29
God, 218–19
The Golden Age (Wright), 41
Goldman Sachs, 68
Good, Irving John, 7

Google, 15, 35, 91, 140, 168, 186, 209
Göring, Hermann, 24
Gottfredson, Linda, 64
Gould, Eric, 194
Grace, Katja, 201–2
gratification, postponed, 80
Greek mythology, 41
Greely, Henry, 109–10

H
Hanson, Robin, xviii
    brain emulation as conditional on
        civilization not collapsing, 10
    bullet-eater, 141, 179
    Cryonics Institute, 214
    emulation of human brain, 139–41
    emulation scenario, 181, 188–89
    emulation technology and rabbit
        population growth, 142–43
    Fermi's paradox as a "great filter," 200
    mankind in a computer simulation,
        45
    materialistic view of life, 139
    Overcoming Bias (blog), 138, 207
    quote, 209
    Singularity materialized through
        emulations, 203
    starvation pressures and emulations,
        150
Hardy, G. H., 107
*Harry Potter and the Methods of
    Rationality* (Yudkowsky), 37–38
Harvard Crimson, 86
Hawking, Stephen, x
Hawks, John, 75
Head Start, 172
health care, 170–71, 180
Heckman, James, 68–69
hedge fund managers, 186
hedge funds, 170, 185–86
Hemingway, Ernest, 92
high school
    boys, 63–64
    football quarterbacks and attractive
        girls, 75

girls, 75, 87
girls, eggs as highly desirable, 88
nurse, 63
Hitler, Adolf, 22
Holmes, Oliver Wendell (Supreme
    Court Justice), 127–28
home buyers, 181
Home Depot, 204
horses, domesticated, 78
house painter, 70
human ancestry, 202
human egg
    advertising, 86–87
    buyers exclusion criteria, 88
    donor compensation, 87
    donors, 86–87
human environmentalists, 152
human goals, 29
human growth, biological limits to, 6
human-level AI. *See also* artificial
    intelligence (AI)
    chess skills, 27
    chess skills, seeks to maximize its, 27
    cognitive powers, wanting to improve
        its, 14
    copies of itself, able to make, 26
    creation of, 15
    destroying our part of the universe,
        could end up, 27
    education, value of education, 193
    Google wrote the code for, 15
    hardware requirements of, minimal, 14
    intelligence explosion after creating,
        15–16, 53, 58
    Intel's program in 2029, xvii
    programmer coding a friendly, 37
    species' survival depends on, 36
    unfriendly, 26
    universe, destroys the, 27
    video recommendations by, 20
    watching porn videos and blackmail-
        ing politicians, 26
    wielding nanotech, 14
human life span, 170
humans aged 70+, 198–99

Hungarian scientists, xii
hunter-gatherers, 76–79, 165, 229–187
"hunters in a farmer's world," 107.
    See also Attention Deficit
    Hyperactivity Disorder (ADHD)
hunting, 75–77
hunting and gathering, 77
hunting strategies, 77
hydrogen bombs, xiii, 8, 57
hyper-genius, 90, 121, 127
hyperlexia, 91
hyperlexic children, 91

I

I, Pencil (Reed), 204–5
immigrant workers, 123, 144
immigration, large-scale, 157
immortality, xv, 46, 176, 180
Imperfect Conceptions (Dikötter), 94
indoor bathing facilities, 166
Industrial Revolution, 77, 81–82, 107,
    128, 142, 165
inequality
    from biological enhancements,
        171–74
    brain-boosting drugs reduces, 125
    brain training reduces, 117
    consumption, 167–69
    eugenics, 125
    intelligence enhancements reduce,
        165
    IQ's impact on, 66
    of Louis XIV's reign, 165
    money as a measure of, 170
Information-based goods, 168
inheritance, 81
innovation race, 204
Intel, xvi–xvii, 5, 12, 17
intelligence. *See also* IQ (intelligence
    quotient)
    additive and non-additive genes and,
        72–73
    biological and machine, xx
    crystallized, 115

enhancement, 90, 113, 115, 119, 121,
    127, 132
enhancements reduce inequality, 165
extraterrestrial, 122, 200–201
fluid, 115–16
general, 44, 64, 114–15, 210
genes for, 16
genetic basis of human, xv, 73, 84
grand theory of, 15
nonbiological, 8–9
in the solar system, 188
technologies that increase human, xvi
von Neumann–level or better, 95
intelligence-augmenting techniques, 113
intelligence-enhancing
    brain implants, 210
    drugs, 101, 155
    genetics, 172
    technologies, 101, 117, 173, 177
intelligence equality, 117
intelligence explosion
    about, 13–16
    accidentally creating, 53
    emulations and, 153
    property rights expectations, 190
    rate of return expectations, 190
    by recursive self-improvement, 14
    savings, cumulative effect on, 190
    threat to mankind's survival from, 121
    ultra-AI, could create, 35
    ultra-AI would emerge from, 187
    unfriendly, 22–28
    wealth expectations, 190
intelligent life-forms, 201
interest rates
    annual, 80
    falling, 81
    low, 80
International Space Station, 105
investment capital, 57–58, 122, 140
investments, long-term, 82
iPod, 205
IQ (intelligence quotient)
    academic success, as excellent predic-
        tor of, 64–65

additive genes explain 60 percent of
the variation in IQ, 72
adopted children studies of, 71–72
adult life, as stable across, 68
age eight, determined by, 67, 96
all children have same academic
potential, 65
Ashkenazi Jews have highest level of
IQ of any population, 96–98
autistics and above-average IQs,
91–92
brain size and, 74
Ceretrophin and six-point increase in
IQ, 106
childhood illness and, 66
cognitive abilities are positively correlated with, 65
cognitive abilities *vs.*, 68
crime and low scores, 70, 74
criminality, poor health, chronic
unemployment and low scores,
85, 155
culture-specific knowledge, measures
something other than, 67
dental care and high scores, 66
digital span and high score, 67
disease burden lowers, 71
economic growth and high scores, 70,
155
economic policies and high scores, 70
economic success and high scores,
68–69
efficient, honest bureaucracies and
high scores, 70
embryo selection to eliminate below-
average IQs, 85
fertility clinics to select for higher
child's IQ, 86
fewer accidents and high scores, 66
"Flynn Effect," 67, 76
genes play a role in correlations of, 66
genetics determines between 50 and
80 percent of, 71
good government and high scores,
70

health, future-orientedness, attractiveness, and lack of criminality
and high scores, 90
high-IQ children, creating extremely,
126
human egg marketing and, 86–87
identical twins study of, 71–72
intelligence, as meaningful measure
of, 73
job performance, as best predictor of,
68
life expectancy and high scores, 66
long-term investments and high
scores, 70
long-term orientation and high
scores, 70
National Football League and high
scores, 90
national growth rate and the nation's
average, 74
neglectful parents and low scores,
71
noncognitive ability and wage variations, 68
parents have little or no effect on
children's, 72
parents select for extremely high-IQ
child, 90
physical attractiveness and high
scores, 66
of poor nations, 70
positive traits, correlated to, 63, 90
poverty and low scores, 72
professions, high IQ, 98
reaction time and, 74
reading enjoyment and high scores,
71
success, best predictor of, 65
super-geniuses are autistic, not all,
91–92
tests are highly *g*-loaded, 64
trust and high scores, 70
wage income and, 74
"win-win" opportunities and high
scores, 70

"IQ War," 68
iterated embryo selection, 98–99

J
James Bond villains, 203
The Jetsons (TV show), 210
Jobs, Steve, 207
Jones, Garett, 64–65

K
Kahneman, Daniel, 205–6
Kasparov, Gary, 4, 132
Kennedy, John F., 124
Khan, Genghis, 22, 24, 77–78
Khrushchev, Nikita, 220
Kling, Arnold, 107
Kool-Aid, 38
Korean study on autism, 91
Krishna (Hindu God), 3
Kurzweil, Ray
    billions of nanobots in our brains will
        record real-time data on how
        our brains work, 11
    computing power, limits to exponen-
        tial growth in, 6
    computing speed, exponential
        improvements in, 4
    Cryonics Institute, 214
    on exponential growth, 1
    *Fantastic Voyage: Live Long Enough to
        Live Forever,* 179
    humans will be fantastically produc-
        tive and earn higher wages,
        189
    investor and Singularity writer, 35
    mankind will colonize the universe at
        maximum speed allowed by the
        laws of physics, 9
    merger of man and machine, 188
    quote, 164
    resources of the solar system, man-
        kind will use significant percent-
        age of, 9
    rocks, reorganization of, x
    Singularity by 2045, 200
    Singularity by a steady merger of man
        and machines, 23
    The Singularity is Near, 3, 207
    Singularity will probably be utopian,
        179
    *Transcend: Nine Steps to Living Well
        Forever,* 179
    ultimate laptop computer operations
        per second, 6
Kurzweilian Merger
    babysitters, earnings of, 134–35
    dangers of, 21
    human brains will provide starter
        software for a Singularity, 8–10
    opportunities to correct mistakes, 22
    property rights expectations, 190
    rate of return expectations, 190
    savings, cumulative effect on, 190
    wealth expectations, 190
Kurzweilian scenario, 189

L
landed aristocracy, 147
landowning nobility, 137
land resale value, 181–82
language processing skills, 64–65
laser rifle, 204
lawn mower, perfect, 24
Legg, Shane, 13
Lesbian feminists, 195
libertarian AI, 41
libertarian government, ideal, 41
libertarianism, 40–41
libertarian ultra-intelligent ruler, 40–41
life span, 180, 212
life span of products, 183
liquid nitrogen preserved body, 139. *See
    also* cryonics
Long Term Capital Management, 185
long-term planning, 79
lottery tickets, IQ, 113
Louis XIV, King, 165, 167
Ludd, Ned, 131
Luddites, 131

Lumosity (brain fitness website), 113, 115
Lumosity games, 113–14

M
Macbeth (Shakespeare), 21, 175
machine intelligence, ix, x, xv, xvii, 13, 21, 119
machine-learning software, xi
Mafia, 217
maggot-free, 166
magic wands, 137
magnetic resonance imaging, 91
Malthus, Thomas, xv, 141
Malthusian emulation society, 149
Malthusian trap, 141–45
Mandarin, 8, 193
Manhattan Project, xii
manic state, 93
mankind's annihilation, 45
mankind's short-term survival prospects, 199
Mao Zedong, 122
marginal costs of production, 168
marijuana, 105
market economy, 132
marriage decision, 83
marriage market, 194
Mars, 138
Martian land, 138
Marxist dictate, 41
"massive embryo selection," 94–95
McCabe, Tom, 90, 163, 215
media content, 182
Medicare costs, 116
Medicare taxes, 157
memes, 80, 97
memory drugs, 108. *See also* cognitive-enhancement drugs
mental ability, general, 64
Methuselah Foundation, 170
Mexicans, millions of illegal, 135
microprocessors, 122
military spending, 180
military threats, 126–28
Milky Way galaxy, 45, 199

minority students, "acceptable" number of, 87
missile technology, 124
modafinil (cognitive-enhancement drug), 104
"modafinil squared" technology, 159
modern man, 77
"A Modest Proposal: Allow Women to Pay for College in Eggs" (Miller), 88
moons of Jupiter, 41
Moore, Gordon, xvi
Moore's Law, xvi 3–5, 8–9, 11, 17, 209
Muehlhauser, Luke, 7
multimillionaires, self-made, 209
murder, 39
musicians, 108

N
nanobots, 10
nanosensors, 127
nanotechnology
    based weapons, 127
    computing hardware, to improve, 8
    to control our emotions and levels of happiness, 8
    free energy from our galaxy stars and, 45
    moons of Jupiter and, 41
    office buildings, to construct large, 181
    restorative, 213
    virtual reality and, 171
nanotech weapons, 197
nanotubes, 4, 17
narcolepsy, 184
narcoleptics, 105
NASA, 217
National Center for Education Statistics (Washington, DC), 172
national defense, 180
National Football League (NFL), 90
natural disasters, 197
natural selection, 171
Nature, 109

Nazi Germany, 23
Netflix, 20
neuroscience, 185
neuroscientists, 13, 17, 203
Newton, Isaac, 91
New York Times, x
NFL. *See* National Football League (NFL)
Nobel science prizes, 96
nonparallel processing, 18
Normans in 1066, 187
North Korea, 187
Norvig, Peter, 35
nuclear war, xi, 197. *See also* thermonuclear war
nuclear weapons, 24, 126
nutritional-supplement regime, 179

O
Obama, Barack (President), 73
obsolescence, 144, 147
The Odyssey (Homer), 61
Omohundro, Steve, 25
Overcoming Bias (blog), 138, 207

P
Pac-Man video game, 209
parallel processing, 18–19
Parkinson's disease, 168–69
Pascal, Blaise, 208
patents and copyrights, 143
people, long term–oriented, 80
person, anonymous, 93
pharmaceutical product development, 183
phonetic pattern of language, 91
pirate maps, 184
placebo effect, 110–11
plagues, 36, 45, 78
plastic surgery, 89
Plath, Sylvia, 92
political correctness, 172
Polizzi, Eric, 5
population groups, 75–76, 96, 173
pornography, hard-core, 38, 195

pornography, Internet, 194
post-Singularity
  civilization, 199–200, 221
  goods, 42
  operating-system world, 41
  pre-Singularity property will have value of, 188
  pre-Singularity will be worthless, 187
  property rights, 56, 188
  race throughout the galaxy, 199
  ultra-AI and chess, 132
  value, 189
  value of education, 192
  value of money, 211
Praetorian Guard, 148
pre-Singularity
  destructive technologies, 201–2
  investments, ultra-AI might obliterate the value of, 187
  property rights, 56, 187–89
  value of money, 211
prisoner-of-war camp, 31
Prisoners' Dilemma
  AI development and, 47–53
  annihilation of mankind, xix
  Chinese militaries and, 48–53
  drug use and risk of schizophrenia or kidney failure, 160–62
  unleaded *vs.* leaded petrol (gas), 57
  US militaries and, 48–53
probe, self-replicating, 199–200
procrastination, 106
production wands, 145
pro-eugenic Chinese, ix
prognosticators, 206–7
property owning, 147
property rights
  economic behavior and, xviii
  post-Singularity, 56
  stable, 82
property rights, pre-Singularity, 187
property rights of bio-humans, 149
Psychology Today, 195
psychotic breakdowns, 120

Q

quantum computing, 5, 17
quantum effects, 4

R

rabbit population, 142
race, star-faring, 200
racial classifications, 76
racial equality, 173
rapture of the nerds, 208
Rattner, Justin, 35
real estate developer, 181–82
real estate development, 188
recessive condition, inherited, 83
Recursive Darkness (horse), 55, 57
Reed, Leonard, 204–5
religious disagreement, 43
reproductive fitness, 76
reproductive success, 75
resale value, 181
residential housing, 181
The Restaurant at the End of the Universe
    (Adams), 150
retirement savings, 175–76
reverse-engineering biology, 203
reversible computing, 17
Ricardian comparative advantage, 136–
    37, 143, 188, 190
Ricardian scenario, 189
Ricardo, David, xvii, 135, 143
rich investors, 144–45
Ritalin (cognitive-enhancement drug),
    xiv, 104–5
Robin. *See* Hanson, Robin
robots, 17, 210
robot soldiers, 24, 53
Roman Empire, 24, 78
Roman Republic, 137–38
Rumsfeld, Donald (former US Secretary
    of Defense), xviii
Russia, 187, 194, 206

S

safety-enhancing products, 179
Salamon, Anna, 44
SAT tests, 66, 162–63

savers gamble on the future, 176
Savoca, Elizabeth, 171
Scientific American
    "Why Women Live Longer," 179
Scottish Mental Survey (1932), 68
search engines, 167
Second Law of Thermodynamics, 27
seed AI
    computer program of roughly
        human-level intelligence capable
        of improving itself, 36
    intelligence explosion, undergoing, 48
    that undergoes an intelligence explo-
        sion, 36
Self-Aware Systems, 25
sequence *F, U*, 115
sexbots, 193–95
sex drives, 195
sex-selective abortions, 194
sexually transmitted diseases, 194
Shakespeare, 21
Shaw, George Bernard, 84
Shulman, Carl, 147, 202
Singularitarians, 215
Singularity
    AI as smart as humans and/or aug-
        menting human intelligence, x
    AI-centered, 209
    AI-induced, 21
    bad, avoid making, 58–59
    bad Singularity-like event, x
    being well-educated will not make
        you materially better off, 192
    believers, 27, 208
    beyond which human affairs can not
        continue, xv
    civilization, xviii
    disabled children and, 212
    emulations, materialized through,
        203
    enabling of computers, 6
    expectations and savings rates,
        186–87
    expectations raise people's uncer-
        tainty about the quality of the
        future, 193

expectations should affect when you
have children, 193
five undisputed facts about, x–xii
by gradual improvement, 178
harm of an unfriendly, 44
hellish, destroys the value of money,
58
humanity will be richer from, 165
intelligence explosion, from an, 203
investments, destroys the value of all,
57
is near, 177, 185, 191, 201, 207
Kurzweil-type, 178
luxuries from, 167
money has no value, 192
money no longer has value, 56
operating-system, 41
pollution as a bad, 56–57
post-Singularity civilization, 199–
200, 221
post-Singularity race throughout the
galaxy, 199
pre-Singularity destructive technologies, 201–2
pre-Singularity property rights will
be respected, 187
pre-Singularity property rights will
have value of post-Singularity,
188–89
pre-Singularity value of money, 211
profit opportunities of, 178
property rights and savings, 187–88
rapid political change, would bring
about, 187
rich, makes everyone extremely, 192
savings and future wealth, 189
savings and investment returns,
188–89
savings rate, plummeting of, 191–92
savings rates with different Singularity scenarios, 189–90
scarcity, might abolish, 187
signposts of the, 177–78
utopian, 56, 58
utopian, will probably be, 179
as utopian or dystopian, 56

value of money, destroys the, 181
watch list, 209
Singularity conference, 92
Singularity Institute for Artificial
Intelligence
to create a "seed AI" designed to
undergo an intelligence explosion to become a friendly ultra-
AI, 14
Michael Vassar as director and former
president, 44
Peter Thiel financial backing of, viii
Yudkowsky founded, 35–36
The Singularity Is Near (Kurzweil), 3,
207
Singularity materialized through
emulations, 203
Singularity movement
intelligence explosion is a barrier to
entry into, 38
Singularity of a black hole, xviii
Singularity University
Larry Page founded, x
slaveholder compensation, 187
slave labor, 137–38
sleep deprivation, 105, 109, 124, 155. *See
also* amphetamines ("speed")
Smith, Adam, 135
Smith College
Adderall, 102–7, 112, 163
amphetamines use, 102
Dean and Adderall-type drugs for
performance-enhancement, 102
student illegal drug use, 101
"study buddy" drugs, 102
survey of illegal cognitive-enhancing
drug use among undergraduates,
103–9
socialists, 41
Social Security taxes, 157
sociopath, 22, 93
sociopathic children, 84
Socrates, 91
Socratic questioning method, 215
soft toilet paper, 166
Soviet Union, xiii, 19, 49, 124, 127, 206

spacecraft, 199
species extinction, 29
Stalin, Joseph, 22, 220
standard of living, 76, 123
Stanovich, Keith, 65–66
StarCraft II (video game), 106
stars "turned of" to conserve energy, 199
Star Trek, 171
starvation pressures, 150
Stewart, Potter (US Supreme Court
    Justice), 38–39
stop signal reaction time, 105
Study of Mathematically Precocious
    Youth, 65
subjective judgment, 39
sub-Saharan Africa, 173
suicide, 92–93
super genius, 90–91, 95
superhuman intelligence, xiv
superintelligence, 21
superintelligence, "alien-like," 122
super-skyscraper, 181
superweapon, 204
surrogate woman, 194
"survival of the richest," 81
surviving children, 82
Swift, Jonathan, 88

T
Tallinn, Jaan, 35, 215
tampons, 166
Tao, Terence, 91–92
tax on emulations, 150
teleportation device, xi
teleportation machine, 138–39
terminal disease, 219
thermonuclear war, 52–53. *See also*
    nuclear war
Thiel, Peter, x, 35, 170, 186, 214
torsion dystonia, 97–98
toxic garbage dumps, 124
trade with extraterrestrials, 122
*Transcend: Nine Steps to Living Well
    Forever* (Kurzweil), 179
transistors, 4
trial-and-error methods, 30

Trident submarine, 23
True Names . . . and Other Dangers
    (Vinge), 36
trust, 70
Turing test, 177
23andMe (testing company), 168–69
2001: A Space Odyssey (movie), 210

U
Ulam, Stanislaw, xv
ultra-AI. *See also* artificial intelligence
    (AI)
    atoms in our solar system, could com-
        pletely rearrange the distribution
        of, 187
    code, made up of extremely complex,
        30
    code, might change its code from
        friendly to non-friendly, 31
    in computer simulation run by a more
        powerful AI, 45–46
    "could never guarantee with "prob-
        ability one" that the cup would
        stay on the table," 28
    free energy supply, will obtain, 27
    friendly, 14, 33, 46, 208
    human destruction because of hyper-
        optimization, 28
    with human-like objectives, 29
    humans don't get a second chance
        once it is created, 30
    indifference towards humanity and
        would kill us, 27
    indifferent to mankind and creation
        of conditions directly in conflict
        with our continued existence, 28
    intelligence explosion and, 31, 35,
        121, 187
    is not designed for friendliness and
        could extinguish humanity, 30,
        36
    lack patience to postpone what might
        turn out to be utopia, 46
    manipulation through humans to win
        its freedom, 32
    martial prowess, 24

military technologies, will discover, 24
morality, sharing our, 29
as more militarily useful than atomic
weapons, 47
power used to stop all AI rivals from
coming into existence, 24
pre-Singularity investments, might
obliterate the value of, 187
progress toward its goals increased by
having additional free energy, 27
rampaging, 23
risks of destroying the world, 49
unfriendly (Devil), 30, 35, 46, 202,
208
unlikely events, will plan against, 28
will command people with hypnosis,
love, or subliminal messages, 33
ultra-intelligence, 40, 44, 47
unfriendly. *See also* friendly
artificial intelligence (AI), 44, 187
intelligence explosion, 22–28
ultra-AI (Devil), 30, 35, 46, 202, 208
University of Kentucky, 103–4
US Civil War, 187
US Department of Defense (DOD), 8
US militaries and Prisoners' Dilemma,
48–53
US weapons policy, xiii

V
vaccines, 166
vacuum tubes, 4
van Gogh, Vincent, 92
Van Sickle, Stephen, 215
Vaseline, 111
Vassar, Michael, 44
venture capitalists, 123, 185–86
video games, 106, 113, 129, 155, 167,
183, 209, 212
Vinge, Vernor, ix, xviii, 36
virtual reality, 42, 139, 150, 171, 181,
210
virtual-reality technology, 183
Virtual World Golf, 167
visual pattern recognition, 105
volcanoes, super, 197

von Neumann, John, xii–xiii, xv, 96,
199–200
von Neumann–level AI, 6

W
wages of cab drivers, 156
Walmart, 204
Washington Post, 172
weapons of mass destruction, 201
website
http://www.lesswrong.com, 37
Weiner, Zach, 151
welfare, 146–47
welfare recipients, 125
What Intelligence Tests Miss (Stanovich),
65–66
white-collar jobs, 194
"Why Women Live Longer" (Scientific
American), 179
widget factory, 57
Wikipedia article, 69, 104, 133
Witten, Ed, 96
Wolfram, Stephen, 35
women, fertile, 78
women past optimal egg age, 88
worker productivity, 132, 140
working memory, 105, 113–16, 212
World of Warcraft (video game), 106, 167
wormhole, 165
Wright, John C., 41
Wrong, Less, 37

Y
Yale Daily News, 86
year 1066, 187
year 1997, 108
year 2000, 35, 87
year 2001, 210
year 2002, 36
year 2004, 181
year 2005, 103, 194
year 2006, 103
year 2007, x, 108
year 2008, 35, 109, 126
year 2010, x
year 2011, 67, 70, 106, 116

year 2012, 11
year 2015, 115
year 2020, 5, 72, 216
year 2021, 108
year 2022, 96
year 2023, 119
year 2025, 8, 36–37, 193
year 2027, 37
year 2029, xvi, xvii, 177
year 2030, xviii, 5, 9, 37, 99
year 2035, 178
year 2045, 9, 37, 175–76, 178, 200, 216
year 2049, xvii, 21
year 2050, 11
year 2080, 37
Yesalis, Charles E., 100
Yew, Lee Kuan, 92
YouTube video, 17–20
Yudkowsky, Eliezer
    biological humans and emulations,
        147
    cryonics as "an ambulance ride to the
        future," 213
    Cryonics Institute, 214
    extrapolation theory, 42

friendliness, preferred approach to, 41
game involving super-smart AI,
    32–33
*Harry Potter and the Methods of Ratio-
    nality,* 37–38
helping people "reprogram" them-
    selves to become more rational,
    37
humanity is at a very unstable point
    in history, 45
quotes of, 21, 34, 55, 83
seed AI will undergo intelligence
    explosion, 36
Singularity as simple but wrong, 41
Singularity from an intelligence
    explosion, 203
Singularity Institute for Artificial
    Intelligence, 35–36, 208
theory of friendly AI, incomplete,
    36–37
Yudkowsky/Hanson debate, 203–5

Z
Zeus (god), 41–42